Statistics in Clinical Vaccine Trials

Jozef Nauta

Statistics
in Clinical
Vaccine Trials

 Springer

Jozef Nauta
Solvay Pharmaceuticals
Global Statistics
C.J. van Houtenlaan 36
1381 CP Weesp
Netherlands
j.nautaekimp@worldonline.nl

ISBN 978-3-642-44191-2   ISBN 978-3-642-14691-6 (eBook)
DOI 10.1007/978-3-642-14691-6
Springer Heidelberg Dordrecht London New York

Cover design: WMXDesign GmbH, Heidelberg

Printed on acid-free paper

Springer is part of Springer Science+Business Media (www.springer.com)

*To my wife Erna Kimp,
and our son Izaak.*

Preface

This book is intended for statisticians working in clinical vaccine development in the pharmaceutical industry, at universities, at national vaccines institutes, etc. Statisticians already involved in clinical vaccine trials may find some interesting new ideas in it, while colleagues who are new to vaccines will be able to familiarize themselves quickly with the statistical methodology.

A good knowledge of statistics is assumed. The reader should be familiar with hypothesis testing, point and confidence interval estimation, likelihood methods, regression, mathematical and statistical notation, etc. A book that would provide the necessary background is: Armitage P., Berry G. and Matthews J.N.S. *Statistical Methods in Medical Research*, 4th edition, Blackwell Science, New York, 2001.

The scope of the book is practical rather than theoretical. Many real-life examples are given, and SAS codes are provided, making application of the methods straightforward. SAS codes are also given for accurate sample size estimation, including codes for the estimation of required sample sizes for equivalence and noninferiority vaccine trials.

The first two chapters are introductions to the immunology of vaccines, and they will provide the reader with the necessary background knowledge. In Chap. 1, the fundamentals of vaccination, the immune system and vaccines are presented. The principle of vaccination is explained, and the major infectious microorganisms are introduced. The primary defence mechanism of microorganisms – antigenic variation – is discussed. A sketch of the immune system is given so that the reader will understand roughly how it works, including the distinction between the innate and the adaptive immune system. The chapter proceeds with a short section on the basics of tumour immunology. An overview of the several types of vaccines for viruses and bacteria, from the first generation live-attenuated vaccines to third generation vaccines such as recombinant vector vaccines, DNA vaccines and virus-like particles vaccines is given. As an example of a parasite vaccine, a summary of the state of affairs of malaria vaccine development is given. Therapeutic vaccines for noninfectious diseases are briefly touched upon. Humoral immunity, the component of the immune system involving antibodies that circulate in the humor, and cellular immunity, the component that provides immunity by action of cells, are explained. Antibody titres and antibody concentrations are introduced, and two standard assays for humoral immunity, the haemagglutination inhibition test and ELISA, are discussed. The distinction between T helper cells and T killer cells and their different roles are explained. A number of assays for cellular immunity are briefly introduced, including the ELISPOT assay.

Chapter 3 is the central one of the book. The four standard statistics to summarize humoral and cellular immunogenicity data are introduced, and in the sections on the statistical analysis of proportions the use of Wilson-type confidence intervals is promoted rather than the more familiar Wald-type intervals. It is explained how exact confidence intervals for the risk difference and the relative risk can be obtained.

In Chap. 4, two types of possible bias for antibody titres are discussed. The first type of bias is due to how antibody titres are defined. An alternative definition is proposed, the mid-value definition. With this definition, the bias is properly corrected. This type of bias is largely of theoretical interest only. That cannot be said of the second type of bias, which is of major practical importance. It occurs when titres above (or below) a certain level are not determined. If this bias is ignored, the geometric mean titre will be underestimated. It is shown how the method of maximum likelihood estimation for censored observation can be applied to eliminate this bias.

Pre-vaccination or baseline antibody levels need not to be zero. Examples of infectious diseases for which this can be the case are tetanus, diphtheria, pertussis and tick borne encephalitis. Imbalance in pre-vaccination state, i.e., a difference in baseline antibody levels between vaccine groups, can complicate the interpretation of a difference in post-vaccination antibody values. A standard approach to this problem is analysis of covariance. But in case of antibody values one of the assumptions underlying this analysis, homoscedasticity, is not met. The larger the baseline value the smaller the standard deviation of the error term. In Chap. 5, a solution to this problem is offered. It is shown that the heteroscedasticity can be modeled. A variance model is derived, and it is demonstrated how this model can be fitted with SAS.

Many vaccine immunogenicity trials are conducted in an equivalence or noninferiority framework. The objective of such trials is to demonstrate that the immunogenicity of an investigational vaccine is comparable or not less than that of a control vaccine. In Chap. 6, the statistical analysis of such trials is explained, both for trials with an antibody response as endpoint and trials with seroprotection or seroconversion as endpoint. The standard analysis of lot consistency data is known to be conservative, but a simple formula is presented which can be used to decide if the lot sample sizes guarantee that the actual type I error rate of the trial is sufficiently close to the nominal level. The chapter is concluded with a discussion of sample size estimation for vaccine equivalence and noninferiority trials, including lot consistency trials. Recommendations are given how to avoid that the statistical power is overestimated.

Chapter 7 considers vaccine field efficacy trial. The aim of a field efficacy trial is to demonstrate that a vaccine protects against infection or disease. First, an overview of the different effects vaccines can produce is given. Next, some critical aspects of such field efficacy trials are discussed. The three most common incidence measures for infection are presented: the attack rate, the infection rate and the force of infection. The statistical analysis of field efficacy trials using these estimators is explained. The chapter then continues with the statistical analysis of recurrent infection data, which is known to be complex. The chapter is concluded with a discussion of sample size estimation for vaccine field efficacy trial. It is shown that the standard method to estimate the sample size for a trial comparing two attack rates and with the aim to demonstrate super efficacy is highly conservative. An SAS code to compute sample sizes for trials comparing two infection rates is presented.

A correlate of protection is an immunological assay that predicts protection against infection. The concept is the topic of Chap. 8. In clinical vaccine, trials correlates of protection are widely used as surrogate endpoints for vaccine efficacy. The function specifying the relationship between log-transformed immunogenicity values and the probability of protection against infection, conditional on exposure to the pathogen, is called the protection curve. It is demonstrated how the parameters of the protection curve can be estimated from challenge study data and vaccine field efficacy data. Also explained is how a threshold of protection can be estimated from the protection curve. The generalizability of estimated protection curves is discussed.

The final chapter, Chap. 9, addresses vaccine safety. To proof the safety of a vaccine is much more difficult than proving its efficacy. Of many vaccines millions of doses are administered, which can bring very rare but serious adverse vaccine events to light. In this chapter, some statistical aspects of vaccine safety are addressed. Vaccine safety surveillance is briefly discussed, and some recent controversies are recalled. The notorious problem of vaccine safety and multiplicity is discussed at great length. Four different methods to handle this problem are presented, including the recently proposed double false discovery method. The performance of the different methods is illustrated with the help of simulation results. The second part of the chapter is dedicated to the analysis of reactogenicity data.

Amsterdam
December 2009

Jos Nauta

Acknowledgement

I would like to acknowledge my colleague and good friend Dr Walter Beyer of the Department of Virology, Erasmus Medical Centre, Rotterdam, the Netherlands, for his generous advice on Chaps. 1 and 2.

Jos Nauta

Contents

Acronyms

AAP	American academy of pediatrics
AIDS	Acquired immune deficiency syndrome
ANCOVA	Analysis of covariance
AOM	Acute otitis media
AR	Attack rate
CBER	Center for biologicals evaluation and research
CDC	Centers for disease control and prevention
CF	Cystic fibrosis
CL	Confidence limit
CLRS	Constrained likelihood ratio statistic
CMI	Cell-mediated immunity
CTL	Cytotoxic T lymphocyte
DNA	Deoxyribonucleic acid
DTP	Diphtheria, tetanus, pertussis
EIA	Enzyme immunoassay
ELISA	Enzyme-linked immunosorbent assay
ELISPOT	Enzyme-linked immunospot
EMA	European medicines agency
EPPT	Events-per-person-time
FDA	United States food and drug administration
FDR	False discovery rate
FWER	Family wise error rate
GBS	Guillain barré syndrome
GM	Geometric mean
GMC	Geometric mean concentration
gMFI	Geometric mean fold increase
gMFR	Geometric mean fold ratio
GMR	Geometric mean ratio
GMT	Geometric mean titre
gp	Glycoprotein-based
GSD	Geometric standard deviation
HA	Haemagglutinin
HAI	Haemagglutination inhibition

HI	Haemagglutination inhibition
Hib	*Haemophilus influenzae* type b
HIV	Human immunodeficiency virus
HPV	Human papillomavirus
ICH	International conference on harmonisation of technical requirements for registration of pharmaceuticals for human use
IFN-γ	Interferon-gamma
i.i.d.	Independent and identically distributed
IL	Interleukin
IRR	Infection rate ratio
IU	Intersection-union
LCL	Lower confidence limit
LL	Log-likelihood
LRS	Likelihood ratio statistic
MedDRA	Medical dictionary for regulatory activities
ML	Maximum likelihood
MMR	Measles, mumps, rubella
MMRV	Measles, mumps, rubella, varicella
NRA	National registration authority (Australia)
NIH	National institutes of health
NK	Natural killer
OPSR	Organization for pharmaceutical safety and research (Japan)
PBMC	Peripheral blood mononuclear cells
RCD	Reverse cumulative distribution
RD	Rate difference
RNA	Ribonucleic acid
RR	Rate ratio
SBA	Serum bactericidal assay
SD	Standard deviation
SE	Standard error
SFDA	China state food and drug administration
SIDS	Sudden infant death syndrome
SPC	Spot-forming cell
TH1	T helper 1
TH2	T helper 2
TOST	Two one-sided tests
UCL	Upper confidence limit
V	Varicella
VAERS	Vaccine adverse event reporting system
VE	Vaccine efficacy
VLP	Virus-like particles

Chapter 1
Basic Concepts of Vaccine Immunology

1.1 Vaccination and Preventing Infectious Diseases

Then in 1798 Edward Jenner (...) published his work. As a young medical student Jenner had heard a milkmaid say: I cannot take the smallpox because I have had cowpox. The cowpox virus resembles smallpox so closely that exposure to cowpox gives immunity to smallpox. (...) Jenner's work with cowpox was a landmark, but not because he was the first to immunize people against smallpox. In China, India, and Persia, different techniques had long since been developed to expose children to smallpox and make them immune, and in Europe at least as early as the 1500s laypeople – not physicians – took material from a pustule of those with a mild case of smallpox and scratched it into the skin of those who had not yet caught the disease. Most people infected this way developed mild cases and became immune.

John M. Barry
The Great Influenza: The Epic Story of the Deadliest Pandemic in History

Vaccines take advantage of the body's ability to learn how to ward off microorganisms. The immune system can recognize and fight of quickly infectious organisms it has encountered before. As an example, consider chickenpox. Chickenpox is a highly contagious infectious disease caused by the varicella zoster virus. First, there are papules, pink or red bumps. These bumps turn into vesicles, fluid-filled blisters. Finally, the vesicles crust over and scab. Clinical symptoms are fever, abdominal pain or loss of appetite, headache, malaise and dry cough. The disease is so contagious that most people get it during their childhood, but those infected are the rest of their life immune to it. Vaccines contain killed or inactivated (parts of) microorganisms. These provoke the immune system in a way that closely mimics the natural immune response to the microorganisms. Vaccination is a less risky way to become immune, because, due to the killing or inactivation of the microorganisms, it does not cause the disease.

Vaccination, together with hygiene, is considered to be the most effective method of preventing infectious diseases. When not prevented, some infectious diseases have proven to be mass killers. Plague, caused by the bacterium *Yersinia pestis*, has been one of the deadliest pandemics in history. The total number of plague deaths worldwide has been estimated at 75 million people, and the disease is thought to have killed almost half of Europe's population. The pandemic arrived in Europe in

J. Nauta, *Statistics in Clinical Vaccine Trials*, DOI 10.1007/978-3-642-14691-6_1,

the 14th century, and it would cast its shadow on the continent for five centuries, with one of the last big outbreaks occurring in Moscow in 1771. (The reader who wants to learn how it was to be trapped in a plague-ridden city should read Giovanni Boccaccio's *The Decameron* (1353) or Daniel Defoe's *A journal of the plague year* (1722)).

The global death toll from the Spanish influenza pandemic (1918–1920), caused by an influenza virus, is assumed to have been more than 30 million.

Malaria is a potentially deadly tropical disease transmitted by a female mosquito when it feeds on blood for her eggs. In Africa, an estimated 2,000 children a day die from the disease, leading in 2006 to a total number of deaths from the disease of almost one million. The Bill and Melinda Gates Foundation is funding efforts to reduce malaria deaths by 2,015, by developing more effective vaccines. The long-term goal of the foundation is to eradicate the disease.

1.2 Microorganisms: Bacteria, Yeasts, Protozoa and Viruses

Microorganisms (also: microbes) are live forms that cannot be seen by the unaided eye, but only by using a light or an electron microscope. The Dutch scientist Anton van Leeuwenhoek (1632–1723) was the first to look at microorganisms through his microscope. Microorganisms that cause disease in a host organism are called *pathogens*. If a microorganism forms a symbiotic relationship with a host organism of a different species and benefits at the expense of that host, it is called a *parasite*.

Bacteria are unicellular organisms surrounded by a cell wall and typically 1–5 μm in length. They have different shapes such as rods, spheres and spirals, and reproduce asexually by simple cell division. The biological branch concerned with the study of bacteria is called *bacteriology*. Examples of serious bacterial diseases are diphtheria, tetanus, pertussis, cholera, pneumococcal disease, tuberculosis, leprosy and syphilis.

Yeasts are unicellular organisms typically larger than bacteria and measuring around 5 μm. Most reproduce asexually, but some also show sexual reproduction under certain conditions. Yeasts are studied within the branch of *mycology*. Diseases caused by yeasts are, among others, thrush and cryptococcosis.

Protozoa are unicellular organisms, more complex and larger than bacteria and yeasts, typically between 10 and 50 μm in diameter. They usually are hermaphroditic and can reproduce both sexually and asexually. Protozoa are responsible for widespread tropical diseases such as malaria, amoebiasis, sleeping sickness and leishmaniasis. The biological branch of *parasitology* includes the study of protozoa and of certain multicellular organisms such as schistozoma and helminths (parasitic worms).

In contrast with bacteria, yeasts and protozoa, which are cellular live forms, *viruses* are too small to form cells (typically 0.05–0.20 μm in diameter). In the environment, they show no metabolism. For replication, a virus needs to intrude a host cell and take over the cell metabolism to produce and release new virus particles. Viruses contain either DNA or RNA as genetic material. DNA viruses include herpes-, adeno-, papova-, hepadna- and poxviruses. RNA viruses include rhino-, polio-, influenza- and rhabdoviruses. Some RNA viruses have an enzyme called reverse transcriptase that allows their viral RNA to be copied as a DNA version (retroviruses). Well-known viral diseases are herpes, hepatitis B and smallpox (DNA viruses), common cold, poliomyelitis, hepatitis A, influenza, rabies (RNA viruses) and human immunodeficiency virus (HIV) (RNA retroviruses). The study of viruses is called *virology*.

1.3   The Immune System

1.3.1   Basics

The immune system can distinguish between nonforeign and foreign (also: self and nonself) molecules and structures. With this ability, it seeks to protect the organism from invading pathogens – by detecting and killing them. The immune system has two essential components, the innate (inborn) or nonspecific and the adaptive or specific immune system.

The *innate immune system* provides an immediate, albeit nonspecific, response to invading pathogens. It is triggered by cells and molecules that recognize certain molecular structures of microorganisms, and it tries to inhibit or control their replication and spread. In vertebrates, one of the first responses of the innate immunity to infection is inflammation, initiated by infected and injured cells that, in response, release certain molecules (histamine, prostaglandins and others). These molecules sensitize pain receptors, widen local blood vessels, and attract certain white blood cells (*neutrophils*) circulating in the blood stream and capable to kill pathogens by ingestion (*phagocytosis*) as a front-line defence. Neutrophils can release even more signalling molecules such as *chemokines* and *cytokines* (among many others: interferon-γ) to recruit other immune cells, including macrophages and natural killer cells. *Macrophages* reside in tissue and also ingest and destroy pathogens. *Natural killer* (NK) cells can detect infected cells (and some tumour cells) and destroy them by a mechanism which is known as *apoptosis*, cell death characterized by protein and DNA degradation and disintegration of the cell. The innate immune system responds to microorganisms in a general way during the early phase of the infection, and it does not confer long-lasting immunity. In vertebrates, the innate immune system actives the adaptive immune system in case pathogens successfully evade this first line of defence.

The *adaptive immune system* has the remarkable ability to improve the recognition of a pathogen, to tailor a response specific to the actual structure of that pathogen, and to memorize that response as preparation for future challenges with the same or a closely similar pathogen. The adaptive immune system activates bone marrow-derived (*B cells*) and thymus-derived cells (*T cells*), leading to humoral and cellular immunity, respectively (see also Chap. 2). In general, B cells make antibodies that attack the pathogens directly, while T cells attack body cells that have been infected by microorganisms or have become cancerous. When activated, B cells secrete *antibodies* in response to *antigens* (from antibody-generating), molecules recognized as nonself. An antigen can be a part of a microorganism, a cancerous structure or a bacterial toxin. The antibodies that are produced are specific to that given antigen. The major role of antibodies is either to mark the invaders for destruction (which, in turn, is effected by other immune cells) or to inactivate (neutralize) them so that they can no longer replicate.

Like B cells, T cells have surface receptors for antigens. T cells can specialize to one of several functions: They may help B cells to secrete antibodies (*T helper cells*), attract and activate macrophages, or destroy infected cells directly (*cytotoxic T cells*, also: *killer cells*). This improved response is retained after the pathogen has been killed (*immunological memory*). It allows the immune system to react faster the next time the pathogen invades the body. This ability is maintained by memory cells which remember specific features of the pathogen encountered and can mount a strong response if that pathogen is detected again.

In vertebrates, the *immune system* is a complex of organs, tissues and cells connected by two separate circulatory systems, the blood stream and the lymphatic system that transports a watery clear fluid called lymph.

In the red bone marrow, a tissue found in the hollow interior of bones, multipotent stem cells differentiate to either red blood cells (*erythrocytes*), or platelets (*thrombocytes*), or white blood cells (*leukocytes*). The latter class is immunologically relevant; leukocytes mature to either *granulocytes* (cells with certain granules in their cytoplasm and a multilobed nucleus, for example the neutrophils mentioned previously) or mononuclear leukocytes, including macrophages and lymphocytes. Natural killer cells, B cells and T cells belong to the lymphocytes. T cell progenitors migrate to the thymus gland, located in the upper chest, where they mature to functional T cells. In the spleen, an organ located in the left abdomen, immune cells are stored and antibody-coated microorganisms circulating in the blood stream are removed. Finally, the lymph nodes store, proliferate and distribute lymphocytes via the lymphatic vessels.

1.3.2 Microbial Clearance

Virus clearance or elimination of a virus infection, involves killing of infected cells by NK cells and cytotoxic T cells, blocking of cell entry or cell-to-cell transmission by neutralizing antibodies, and phagocytosis by macrophages.

The major process of bacterial clearance is phagocytosis. Pathogenic bacteria have three means of defence against it. The first defence is the cell capsule, a layer outside the cell wall that protects bacteria from contact with macrophages and other phagocytes. The second defence is the cell wall, which acts as a barrier to microbicidal activity. The third defence is the secretion of *exotoxins*, poisonous substances that damage phagocytes and local tissues and, once circulating in the blood stream, remote organs. Frequently, exotoxins (and not the bacteria themselves) are the cause of serious morbidity of an infected organism. Most cell capsules and exotoxins are antigenic, meaning that antibodies can block their effects.

Protozoan clearance is exceptionally difficult. Immunity is usually limited to keeping the parasite density down. Malaria clearance, for example, involves phagocytosis of parasitized red blood cells by macrophages and antibodies. During the brief liver stage of the malaria parasites, immunity can be induced by cytotoxic T cells.

1.3.3 Active and Passive Protection from Infectious Diseases

The immune system can quickly recognize and fight off infectious organisms it has encountered before. Measles is a highly contagious infectious childhood disease caused by the measles virus and transmitted via the respiratory route. Infected children become immune to it for the rest of their life. This is called *naturally acquired active immunity*. Because newborn infants are immunogically naive (no prior exposure to microorganisms), they would be particularly vulnerable to infection. Fortunately, during pregnancy, antibodies are passively transferred across the placenta from mother to foetus (*maternal immunity*). This type of immunity is called *naturally acquired passive immunity*. Depending on the half-life time of these passively transferred antibodies, maternal immunity is usually short-term, lasting from a few days up to several months.

1.3.4 Antigenic Variation

While measles does usually not attack an individual twice in lifetime due to naturally acquired active immunity, some other pathogens try to trick the immunological memory by various mechanisms. One is an adaptation process called antigenic variation: small alterations of the molecular composition of antigens of the surface of microorganisms to become immunologically distinct from the original strain. (A *strain* is a subset of a species differing from other members of the same species by some minor but identifiable change.) *Antigenic variation* can occur either due to gene mutation, gene recombination or gene switching. Antigenic variation can occur very slowly or very rapidly. For example, the poliovirus, the measles virus and the yellow fever virus have not changed significantly since vaccines against them were first developed, and these vaccines therefore offer lifelong protection. Examples for rapidly evolving viruses are HIV and the influenza virus. Rapid antigenic variation is an important cause of vaccine failure.

A *serotype* is a variant of a microorganism in which the antigenic variations are to such a degree that it is no longer detected by antibodies directed to other members of that microorganism. For example, of the bacterium *Pseudomonas aeruginosa*, more than sixteen serotypes are known, of the hepatitis B virus four major serotypes have been identified, and of the rhinovirus, cause of the common cold, there are so many serotypes (more than 100) that many people suffer from common cold several times every winter – each time caused by a member of a different serotype. In case of influenza, antigenic variation is called *antigenic drift*, which is the process of mutations in the virus surface proteins haemagglutinin and neuraminidase. This drift is so rapid that the composition of influenza vaccines has to be changed almost every year. Antigenic drift should not be confused with *antigenic shift*, the process at which two different strains of an influenza virus combine to form a new antigenic subtype, for which the immune system of the host population is naive and which makes it extremely dangerous because it can lead to pandemic outbreaks.

1.3.5 Tumour Immunology

There is considerable evidence that many tumours are eliminated by the immune system at a very early stage, before they become evident (*immune surveillance*). The immune response to a tumour is very complex and not yet fully understood, and it depends on many factors, notably the type of tumour. Natural killer cells and cytotoxic T cells play an important role in tumour control. Another important feature is a high rate of apoptosis of tumour cells. If the rate is too high and there are too many apoptotic cells to be phagocytosed by macrophages, then some of these cells release protein fragments which activate NK and other interferon-γ producing cells. Interferon-γ has multiple roles in immune surveillance, one making tumour cells more readily recognizable by anti-tumour T cells. Tumour cells may express tumour-associated antigens, which are proteins found on the surface of tumour cells at a higher level than of normal cells. The immune system recognizes these antigens as foreign, and reacts by destroying the tumour cells by T cells. The antigen expression can have several causes, one being infection by an oncogenic (cancer-causing) virus. Five forms of cancer are, up to now, known to be caused by viruses: Burkitt's lymphoma, nasopharyngeal carcinoma, Kaposi's sarcoma, hepatocarcinoma and cervical cancer. The latter disease is caused by the human papillomavirus (HPV), the only oncogenic virus for which at present a vaccine is available.

1.4 Prevention of Infectious Diseases by Vaccination

The word vaccination (Latin: *vacca*-cow) was first used by the British physician Edward Jenner (1749–1823) who searched for a prevention of smallpox, a widespread disease localized in small blood vessels of the skin, mouth and throat, causing a maculopapular rash and fluid-filled blisters and often resulting in disfigurement, blindness and death. In 1798, Jenner published his *An inquiry into the causes and effects of the Variolae Vaccinae, a disease discovered in some of the western counties of England, particularly Gloucestershire, and known by the name of the cow-pox*. He reported how he, two years earlier, had taken the fluid from a cowpox pustule on a dairymaid's hand and inoculated an eight-year-old boy. Six weeks later, he exposed the boy to smallpox, but the boy did not develop any symptoms of smallpox disease. Today, the virological background of Jenner's successful intervention is understood: variola virus, the cause of smallpox, and cowpox virus, the cause of a mild veterinary disease with only innocent symptoms in men, are quite similar DNA viruses belonging to the same viral genus orthopoxvirus. Unintendedly, dairymaids were often exposed to, and infected by cowpox virus during milking. Consequently, they developed immunity which also protected against the smallpox virus (cross-protection). Previously, this type of immunity was called naturally acquired active immunity. By intended inoculation with cowpox virus, Jenner had the eight-year-old boy actually achieve artificially acquired active immunity – the aim of any vaccination. The year 1996 marked the two hundredth anniversary of Jenner's experiment. After large-scale vaccination campaigns throughout the nineteenth and twentieth century using vaccinia virus, another member of the same viral genus, the World Health Organization in 1979 certified the eradication of smallpox. To this day, smallpox is the only human infectious disease that has been completely eradicated.

Among the pioneers of vaccinology were the French chemist Louis Pasteur (1822–1895), who developed a vaccine for rabies, and the German Heinrich Hermann Robert Koch (1843–1910), who isolated *Bacillus anthracis*, *Vibrio cholerae* and *Mycobacterium tuberculosis*, a discovery for which he in 1905 was awarded the Nobel Prize. Koch also developed criteria to establish, or refute, the causative relationship between a given microorganism and a given disease (*Koch's Postulates*). This was, and is, essential for vaccine development. First, one has to prove that a given microorganism is really the cause of a given clinical disease, and then one can include that microorganism in a vaccine to protect people from that disease. The causative relationship between a microbe and a disease is not always self-evident. In the first decades of the twentieth century it was widely believed that the cause of influenza was the bacterium *Haemophilus influenzae*, because it was often isolated during influenza epidemics. Only when in the 1930s influenza viruses were discovered and proven, by Koch's postulates, to be the real cause of influenza, the way was opened to develop effective vaccines against that disease. A vaccine containing *H. influenzae* would not at all protect from influenza.

Most vaccines contain *attenuated* (weakened) or inactivated microorganisms. Ideally, they provoke the adaptive immune system in a way that closely mimics the immune response to the natural pathogenic microorganisms. Vaccination is a less risky way to become immune, because, due to the attenuation or inactivation of the microorganisms, a vaccine does not cause the disease associated with the natural microorganism. Yet, naive B and T cells are activated as if an infection had occurred, leading to long-lived memory cells, which come into action after eventual exposure with the natural microorganism.

1.4.1 Viral and Bacterial Vaccines Currently in Use

Live attenuated vaccines contain living viruses or bacteria of which the genetic material has been altered so they cannot cause disease. The classical way of attenuation is achieved by growing the microorganisms over and over again under special laboratory conditions. This *passaging* process deteriorates the disease-causing ability of the microorganisms. The weakened viruses and bacteria still can infect the host, and thus stimulate an immune response, but they can rarely cause disease. However, in certain immune-compromised patients, even attenuated microorganisms may be dangerous so that manifest immune-suppression can be a contra-indication for live vaccines.

An example of a live attenuated vaccine is the RIX4414 human rotavirus. Rotavirus infection is the leading cause of potentially fatal dehydrating diarrhoea in children. The parent strain RIX4414 was isolated from a stool of a 15-month-old child with rotavirus diarrhoea and attenuated by tissue culture passaging. Other examples of diseases for which vaccines are produced from live attenuated microorganisms are the viral diseases measles, rubella and mumps, polio, yellow fever and influenza (an intranasal vaccine), and the bacterial diseases pertussis (whooping cough) and tuberculosis. In general, live attenuated vaccines are considered to be very immunogenic. To maintain their potency, they require special storage such as refrigerating and maintaining a cold chain. There is always a remote possibility that the attenuated bacteria or viruses mutate and become virulent (infectious).

In contrast, *inactivated vaccines* contain microorganisms whose DNA or RNA was first inactivated, so that they are 'dead' and cannot replicate and cause an infection any more. Therefore, these vaccines are also safe in immune-compromised patients. Inactivation is usually achieved with heat or chemicals, such as formaldehyde or formalin, or radiation. There are several types of inactivated vaccines.

Whole inactivated vaccines are composed of entire viruses or bacteria. They are generally quite immunogenic. However, they are often also quite reactogenic, which means that vaccinees may frequently suffer from local vaccine reactions at the site of vaccination (e.g., redness, itching, pain) or even from systemic vaccine reactions such as headache and fever. Fortunately, these reactions are usually benign, mild and transitory, and only last from hours to a few days. Whole vaccines have been developed for prophylaxis of, amongst others, pertussis (bacterial), cholera (bacterial) and influenza (viral).

Component vaccines do not contain whole microorganisms but preferably only those parts which have proven to stimulate the immune response most. The advantage of this approach is that other parts of the microorganism in question, which do not contribute to a relevant immune response but may cause unwanted vaccine reactions, can be removed (vaccine purification). Thus, component vaccines are usually less reactogenic than whole vaccines. Simple component vaccines are the split vaccines, which result after the treatment with membrane-dissolving liquids likesuch as ether. More sophisticated, subunit vaccines are produced using biological or genetic techniques. They essentially consist of a limited number of defined molecules, which can be found on the surface of microorganisms. Their vaccine reactogenicity is thereby further decreased. A disadvantage can be that isolated antigens may not stimulate the immune system as well as whole microorganisms. To overcome this problem, *virus-like particles* (VLP) *vaccines* and *liposomal vaccines* have been developed. Virus-like particles are particles that spontaneously assemble from viral surface proteins in the absence of other viral components. They mimic the structure of authentic spherical virus particles and they are believed to be more readily recognized by the immune system. In liposomal vaccines, the immunogenic subunits are incorporated into small vesicles sized as viruses ($0.1–0.2\,\mu m$) and made of amphiphilic chemical compounds such as phospholipids (main components of biological membranes). Examples of component vaccines are *Haemophilus influenzae* type b (Hib) vaccines, hepatitis A and B vaccines, pneumoccocal vaccines, and, again, influenza vaccines. The current generation of HPV vaccines are virus-like particles vaccines.

Another approach to increase the immunogenicity of inactivated vaccines is the use of *adjuvants*. These are agents that, by different mechanisms, augment the immune response against antigens. A potent adjuvant which has been used for over 50 years is aluminium hydroxide. In recent years, a number of new adjuvants have been developed: MF59 (an oil-in-water emulsion), MPL (a chemically modified derivative of lipopolysaccharide) and CpG 7909 (a synthetic nucleotide). Adjuvanted vaccines tend to more enhanced reactogenicity, i.e., they lead to higher incidences of local and systemic reactions. Some known adjuvants are therefore not suitable for human use (possibly still for veterinary use), for example Freund's complete adjuvant (heat-killed *Mycobacterium tuberculosis* emulsified in mineral oil). It is very effective to enhance both humoral and cellular immunity, but has been found to produce skin ulceration, necrosis and muscle lesion when administered as intramuscular injection. Other potential safety concerns of adjuvanted vaccines are immune-mediated adverse events (e.g., anaphylaxis or arthritis) or chemical toxicity.

The immune system of infants and young children has difficulties to recognize those bacteria which have outer coats that disguise antigens. A notorious example is the bacterium *Streptococcus pneumoniae*. *Conjugate vaccines* may overcome this problem. While an adjuvanted vaccine consists of a physical mixture of vaccine and adjuvant, in a conjugate vaccine the microbial antigens are chemically bound to certain proteins or toxins (the carrier proteins), with the effect that recognition by the juvenile immune system is increased. This technique is used for, among else, Hib and pneumococcal vaccines.

Certain bacteria produce exotoxins capable of causing disease. Diphtheria is a bacterial disease, first described by Hippocrates (ca. 460–377 B.C.). Epidemics of diphtheria swept Europe in the seventeenth century and the American colonies in the eighteenth century. The causative bacterium is *Corynebacterium diphtheriae*, which produces diphtheria toxin. This toxin can be deprived of its toxic properties by inactivation with heat or chemicals, but it still carries its immunogenic properties; it is then called a toxoid and can be used for diphtheria *toxoid vaccine*. Another example is the tetanus vaccine containing the toxoid of the bacterium *Clostridium tetani*.

Diphtheria and tetanus toxoid vaccines are often given to infants in combination with a vaccine for pertussis. This combination is known as DTP vaccine.

DTP vaccine is an example of a *combination vaccine*, which intends to prevent a number of different diseases, or one disease caused by different strains or different serotypes of the same species, such as the seasonal influenza vaccines which currently contain antigens of three virus (sub)types: a B-strain, an A-H1N1 strain and an A-H3N2 strain. This is an example for a trivalent vaccine (including three strains). Pneumococcal vaccines are currently available as 7-valent to even 23-valent vaccines. In contrast, a monovalent vaccine is intended to prevent one specific disease only caused by one defined microorganism, for example the hepatitis B vaccine.

1.4.2   Routes of Administration

Licensed vaccines differ with respect to the route of administration. This is not only a question of comfort for the vaccinee, but also depends on the exact types and location of immune cells to which the vaccine is offered to achieve the optimal prophylactic effect. *Injectable vaccines* are usually given *subcutaneously* (into the fat layer between skin and muscle) or *intramuscularly* (directly into a muscle). Preferred vaccination sites are the deltoid region of the arm in adults and elderly, and the thigh in newborns and infants. Some vaccines – hepatitis B vaccines for example – can also be administered intramuscularly in the buttock. An alternative to subcutaneous/intramuscular injection is *intradermal vaccination*, directly into the dermis. Intradermal vaccination is successfully used for rabies and hepatitis B. In case of influenza, it reduces the dose needed to be given. This route could thus increase the number of available doses of vaccine, which can be relevant in case of an influenza pandemic. Vaccination by injection is often felt to be uncomfortable by vaccinees, it usually needs some formal medical training to administer it, and it carries the risk of needle prick accidents with contaminated blood.

An alternative is administration by the *oral route*, since the 1950s used for the live attenuated polio vaccine: some droplets of vaccine-containing liquid on a lump of sugar to be swallowed. This route builds up a strong local immunity in the intestines, the site of poliovirus entry. Obvious advantages are the increased ease and acceptance of vaccination and the absence of the risk of blood contamination.

A third option is intranasal vaccine administration, preferably used for respiratory pathogens. *Intranasal vaccines* are dropped or sprayed into the cavity of the nose. Advantages are, again, the ease of administration (in particular for childhood vaccines), the direct reach of the respiratory compartment, and hence induction of local protective immunity at the primary site of pathogen entry.

1.4.3 Malaria Vaccines

Malaria is an example of a protozoan disease. The most serious forms of the disease are caused by the parasites *Plasmodium falciparum* and *Plasmodium vivax*. The parasites are transmitted by the female Anopheles mosquito. Sporozoites (from *sporos*, seed) of the parasites are injected in the bitten person. In the liver, sporozoites develop into blood-stage parasites which then reach red blood cells. There are three types of malaria vaccines in development: pre-erythrocytic vaccines, blood-stage vaccines and transmission-blocking vaccines. Pre-erythrocytic vaccines target the sporozoites and the liver life forms. If fully effective, they would prevent blood-stage infection. In practice, they will be only partially effective, but they may reduce the *parasite density* (density of malaria parasites in the peripheral blood) in the initial blood-stage of the disease. Blood-stage vaccines try to inhibit parasite replication by binding to the antigens on the surface of infected red blood cells. These vaccines also may reduce parasite density to a level that prevents development of clinical disease. Transmission-blocking vaccines try to prevent transmission of the parasite to humans rather than preventing infection. This is attempted by trying to induce antibodies that act against the sexual stages of the parasite, to prevent it from becoming sexually mature.

1.4.4 Experimental Prophylactic and Therapeutic Vaccines

Some recent developments in vaccine research, still in an experimental stage in animal models, are recombinant vector vaccines and DNA vaccines. *Recombinant vector vaccines* are vaccines created by recombinant DNA technology. The pathogen's DNA is inserted into a suitable virus or bacterium that transports the DNA into healthy body cells where the foreign DNA is read. Consequently, foreign proteins are synthesized and released, which act as antigens stimulating an immune response. Similarly, *DNA vaccines* are made of plasmids, circular pieces of bacterial DNA with incorporated genetic information to produce an antigen of a pathogen. When the vaccine DNA is brought into suitable body cells, the antigen is expressed, and the immune system can respond to it. The advantage of DNA vaccines is that no outer source of protein antigen is needed. Serious safety concerns will have to be addressed before these experimental approaches can be tested in man.

The vaccines discussed so far were all *prophylactic vaccines*, i.e., intended to prevent infection. A fairly recent development is the emergence of *therapeutic vaccines*, not given with the intention to prevent but to treat. The targeted diseases need not to be infectious. Therapeutic tumour vaccines, for example, are aimed at tumour forms that the immune system cannot destroy. The hope is to stimulate the immune system in such a way that the enhanced immune response is able to kill the tumour cells. Therapeutic tumour vaccines are being developed for acute myelogenous leukaemia, breast cancer, chronic myeloid leukaemia, colorectal cancer, oesophageal cancer, head and neck cancer, liver and lung cancers, melanoma, nonHodgkin lymphoma, and ovarian, pancreatic and prostate cancers. Other examples of therapeutic vaccines being developed are addiction vaccines for cocaine and nicotine abuse. Nicotine is made of small molecules that are able to pass the blood–brain barrier, a filter to protect the brain from dangerous substances. One vaccine in development has the effect that the subject develop antibodies to nicotine, so that when they smoke, the antibodies attach to the nicotine and make the resulting molecule too big to pass the blood–brain barrier, so that smoking stops being pleasurable. Another nicotine vaccine in development leads to the production of antibodies that block the receptor that is involved in smoking addiction. Therapeutic vaccines are also being tested for hyperlipidaemia, hypertension, multiple sclerosis, rheumatoid arthritis and Parkinson's disease.

Chapter 2
Humoral and Cellular Immunity

2.1 Humoral Immunity

When the adaptive immune system is activated by the innate immune system, the *humoral immune response* (also: *antibody-mediated immune response*) triggers specific B cells to develop into plasma cells. These plasma cells then secrete large amounts of antibodies. Antibodies circulate in the lymph and the blood streams. (Hence the name: humoral immunity. *Humoral* comes from the Greek *chymos*, a key concept in ancient Greek medicine. In this view, people were made out of four fluids: blood, black bile, yellow bile and mucus (phlegma). Being healthy meant that the four humors were balanced. Having too much of a humor meant unbalance resulting in illness.) The more general term for antibody is *immunoglobulin*, a group of proteins. There are five different antibody classes: IgG, IgM, IgA, IgE and IgD. The first three, IgG, IgM and IgA, are involved in defence against viruses, bacteria and toxins. IgE is involved in allergies and defence against parasites. IgD has no apparent role in defence. The primary humoral immune response is usually weak and transient, and has a major IgM component. The secondary humoral response is stronger and more sustained and has a major IgG component.

Antibodies attack the invading pathogens. Different antibodies can have different functions. One function is to bind to the antigens and mark the pathogens for destruction by phagocytes, which are cells that phagocytose (ingest) harmful microorganism and dead or dying cells. Some antibodies, when bound to antigens, activate the *complement*, serum proteins able to destroy pathogens or to induce the destruction of pathogens. These antibodies are called *complement-mediated antibodies*. *Neutralizing antibodies* are antibodies that bind to antigens so that the antigen can no longer recognize host cells, and infection of the cells is inhibited. For example, in case of a virus, neutralizing antibodies bind to viral antigens and prevent the virus from attachment to host cell receptors.

It is good practice to state the antigen against which the antibody was produced: anti-HA antibody, anti-tetanus antibody, anti-HPV antibody, etc.

J. Nauta, *Statistics in Clinical Vaccine Trials*, DOI 10.1007/978-3-642-14691-6_2,
© Springer-Verlag Berlin Heidelberg 2011

2.1.1 Antibody Titres and Antibody Concentrations

Antibody levels in serum samples are measured either as antibody titres or as antibody concentrations. An *antibody titre* is a measure of the antibody amount in a serum sample, expressed as the reciprocal of the highest dilution of the sample that still gives (or still does not give) a certain assay read-out. To determine the antibody titre, a serum sample is *serially* (stepwise) *diluted*. The *dilution factor* is the final volume divided by the initial volume of the solution being diluted. Usually, the dilution factor at each step is constant. Often used dilution factors are 2, 5 and 10. In this book, the *starting dilution* will be denoted by $1:D$. A starting dilution of 1:8 and a dilution factor of 2 will result in the following two-fold serial dilutions: 1:8, 1:16, 1:32, 1:64, 1:128 and so on. To each dilution, a standard amount of antigen is added. An assay (test) is performed which gives a specified read-out either when antibodies against the antigen are detected or, depending on the test, when no antibodies are detected. The higher the amount of antibody in the serum sample, the higher the dilutions at which the assay read-out occurs (or no longer occurs). Suppose that the assay read-out occurs for the dilutions 1:8, 1:16 and 1:32, but not for the dilutions 1:64, 1:128, etc. The antibody titre is the reciprocal of the highest dilution at which the read-out did occur, 32 in the example. If the assay read-out does not occur at the starting dilution – indicating a very low amount of antibodies, below the detection limit of the assay – then often the antibody titre for the sample is set to $D/2$, half of the starting dilution. By definition, antibody titres are dimensionless.

Antibody concentrations measure the amount of antibody-specific protein per millilitre serum, expressed either as micrograms of protein per millilitre (μg/ml) or as units per millilitre (U/ml). (A unit is an arbitrary amount of a substance agreed upon by scientists.) The measurement of antibody concentrations can usually done on a single serum sample rather than on a range of serum dilutions.

2.1.2 Two Assays for Humoral Immunity

To give the reader an idea of how antibody levels in serum samples are determined, below two standard assays for humoral immunity are discussed, the haemagglutination inhibition test involving serum dilutions, and the enzyme-linked immunosorbent assay involving a single serum.

Some viruses – influenza, measles and rubella, amongst others – carry on their surface a protein called *haemagglutinin* (HA). When mixed with erythrocytes (red blood cells) in an appropriate ratio, it causes the blood cells clump together (agglutinate). This is called *haemagglutination*. Anti-HA antibodies can inhibit (prevent) this reaction. This effect is the basis for the *haemagglutination inhibition* (HI, also HAI) *test*, an assay to determine antibody titres against viral haemagglutinin. First, serial dilutions of the antibody-containing serum are allowed to react with a constant amount of *antigen* (virus). In the starting dilution and the lower dilutions, the amount of antibody is larger than the amount of antigen, which means that all virus particles are bound by antibody. At a certain dilution, the antibody amount becomes smaller than the antigen amount, which means that free, unbound virus remains. This free antigen is then detected by the second part of the test: to all dilutions, a defined amount of erythrocytes is added. In the lower dilutions, where all antigen is bound by antibody, the erythrocytes freely sink to the lowest point of the test tube or well and form a red spot there (no haemagglutination). In higher dilutions, where there is so less antibody that free virus remains, this virus binds to erythrocytes, which then form a wide layer in the test tube (haemagglutination). The reciprocal of the last dilution where haemagglutination is still inhibited (i.e., where haemagglutination does not occur) is the antibody titre.

The *enzyme-linked immunosorbent assay* (ELISA), also called *enzyme immunoassay* (EIA), is another assay to detect the presence of antibodies in a serum sample. Many variants of the test exist, and here only the basic principle will be explained. In simple terms, a defined amount of antigen is bound to a solid-phase surface, usually the plastic of the wells of a microtitre plate. Then a serum sample with an unknown amount of antigen-specific antibody is added and allowed to react. If antibody is present, it will bind to the fixed antigen. Consequently, the serum (with unbound antibody, if any) is washed away, while the fixed antigen–antibody complexes remain on the solid-phase surface. They are detected by adding a solution of antibodies against human immunoglobulin, which have previously been prepared in animals and chemically linked to an enzyme. The fixed complexes consist of three components: the test antigen, the antibody of unknown amount from the serum specimen, and the enzyme-labelled secondary test antibody against the serum antibody. A substrate to the enzyme is added, which is split by the enzyme, if present. One of the released splitting products can give a detectable signal, a certain colour, for example. Only if the three-component complex is present (i.e., if there has been antibody in the serum specimen), this signal will occur. The strength of the signal is a measure of the amount of serum antibody.

The first-generation ELISA use chromogenic substrates, which release colour molecules after enzymatic reaction. By a spectrophotometer, the intensity of the colour in the solution (or the amount of light absorbed by the solution) can be determined (optical density). The antibody concentration is determined by comparing the optical density of the serum sample with an optical density curve constructed with the help of a standard sample.

In a fluorescence ELISA, which has a higher sensitivity than a colour-releasing ELISA, the signal is given by fluorescent molecules, whose amount can be measured by a spectrofluorometer.

2.2 Cellular Immunity

Cellular immunity (also: *cell-mediated immunity* (CMI)) is an adaptive immune response that is primarily meditated by thymus-derived small lymphocytes, which are known as T cells. Here, two types of T cells are considered: T helper cells and

T killer cells. *T helper cells* are particularly important because they maximize the capabilities of the immune system. They do not destroy infected cells or pathogens, but they activate and direct other immune cells to do so. Hence their name: T helper cells. The major roles of T helper cells are to stimulate B cells to secrete antibodies, to activate phagocytes, to activate T killer cells and to enhance the activity of natural killer (NK) cells. Another term for T helper cells is CD4+ T cells (CD4 positive T cells), because they express the surface protein CD4. T helper cells are subdivided on the basis of the cytokines they secrete after encountering a pathogen. *T Helper 1 cells* (TH1 cells) secrete many different types of cytokines, the principal being interferon-γ (IFN-γ), interleukin-2 (IL-2) and interleukin-12 (IL-12). IFN-γ has many effects including activation of macrophages to deal with intracellular bacteria and parasites. IL-2 stimulates the maturation of killer T cells and enhances the cytotoxicity of NK cells. IL-12 induces the secretion of INF-γ. The principal cytokines secreted by *T Helper 2 cells* (TH2 cells) are interleukin-4 (IL-4) and interleukin-5 (IL-5) for helping B cells. An infection with the human immunodeficiency virus (HIV) demonstrates the importance of helper T cells. The virus infects CD4+ T cells. During an HIV infection, the number of CD4+ T cells drops, leading to the disease known as the acquired immune deficiency syndrome (AIDS).

The major function of *T killer cells* is cytotoxicity to recognize and destroy cells infected by viruses, but they also play a role in the defence against intracellular bacteria and certain types of cancers. Intracellular pathogens are usually not detected by macrophages and antibodies, and clearance of infection depends upon elimination of infected cells by cytotoxic lymphocytes. T killer cells are specific, in the sense that they recognize specific antigens. Alternative terms for T killer cells are CD8+ T cells (CD8 positive T cells), cytotoxic T cells and CTLs (cytotoxic T lymphocytes). CD8+ T cells secrete INF-γ and the inflammatory cytokine tumour necrosis factor (TNF).

2.2.1 Assays for Cellular Immunity

Most assays for cellular immunity are based on cytokine secretion, as marker of T cell response. A wide variety of assays exists, but the most used one is the enzyme-linked immunospot (ELISPOT) assay, which was originally developed as a method to determine the number of B cells secreting antibodies. Later, the method was adapted to determine the number of T cells secreting cytokines. ELISPOT assays are performed in microtitre plates coated with the relevant antigen. *Peripheral blood mononuclear cells* (PBMCs) are added to it and then incubated. (PBMCs are white blood cells such as lymphocytes and monocytes). When the cells are secreting the specific cytokine, discrete coloured spots are formed, which can be counted. One of the most popular of this type of assays to evaluate cellular immune responses is the INF-γ ELISPOT assay, an assay for CTL activity. Results are expressed as *spot-forming cells* (SPCs) per million peripheral blood mononuclear cells (SPC/10^6 PMBC). Other types of the assay are the IL-2 ELISPOT assay, the IL-4 ELISPOT assay, etc.

The *fluorospot assay* is a modification of the ELISPOT assay and is based on using multiple fluoroscent anticytokines, which makes it possible to spot two cytokines in the same assay.

Other assays that can quantitate the number of antigen-specific T cells are the intracellular cytokine assay and the tetramer assay.

Flow cytometry uses the principles of light scattering and emission of fluorochrome molecules to count cells. Cells are labelled with a fluorochrome, a fluorescent dye used to stain biological specimens. A solution with cells is injected into the flow cytometer, and the cells are then forced into a stream of single cells by means of hydrodynamic focusing. When the cells intercept light from a source, usually a laser, they scatter light and fluorochromes are realized. Energy is released as a photon of light with specific spectral properties unique to the fluorochrome.

Chapter 3
Standard Statistical Methods for the Analysis of Immunogenicity Data

3.1 Introduction

There is an ancient proverb, popularized by Spanish novelist Cervantes in his *Don Quixote* (1605), that says that the proof of the pudding is in the eating. Putting it figuratively, ideas and theories should be judged by testing them. For vaccines the test is the field efficacy trial. A group of disease-free subjects are randomized to be vaccinated with either the investigational vaccine or a placebo vaccine. The subjects are then followed-up, to see how many cases of the disease occur in the two arms of the trial. If in the investigational arm the number of cases is significantly lower than in the placebo group, this is considered to be proof that the investigational vaccine protects from infection. Vaccine field efficacy trials, however, have a notorious reputation among vaccine researchers. They are extremely if not prohibitively costly, as they usually require large sample sizes and a lengthy follow-up. If during the surveillance period the attack rate of the infection and thus the number of cases is low, the period has to be extended, meaning even higher costs. Many vaccine field efficacy trials have been negative as the result of imperfect case finding. Further, placebo-controlled vaccine trials in elderly are considered unethical.

A popular alternative to vaccine field efficacy trials are *vaccine immunogenicity trials*. In such trials, the primary endpoint is a humoral or a cellular immunity measurement which is thought to be a correlate of protection from infection. Vaccine immunogenicity trials are usually much smaller and require often only a short follow-up, which makes them less costly than field efficacy trials. The key to the popularity of vaccine immunogenicity trials is that registration authorities such as the United States Food and Drug Administration (FDA), the European Medicines Agency (EMA), Japan's Organization for Pharmaceutical Safety and Research (OPSR), the China State Food and Drug Administration (SFDA) or the National Registration Authority (NRA) of Australia for example, all accept the results of vaccine immunogenicity trials in support of licensure of new vaccines. Indeed, many registration authorities license a vaccine solely on the basis of immunogenicity data, on the condition that the primary immunogenicity measurement is an established correlate of protection.

J. Nauta, *Statistics in Clinical Vaccine Trials*, DOI 10.1007/978-3-642-14691-6_3,
© Springer-Verlag Berlin Heidelberg 2011

Until recently vaccine immunogenicity trials typically focussed on the humoral immune response, i.e., on serum antibody levels. Today, many papers on vaccine immunogenicity report also on cellular immunity. Nevertheless, cellular immunity in vaccine trials is still largely in the investigational phase.

There are four standard statistics to summarize humoral and cellular immunity data: the geometric mean response, the geometric mean fold increase, the seroprotection rate and the seroconversion rate. Two of these statistics, the geometric mean response and the seroprotection rate, quantify absolute immunogenicity values, while the other two, the geometric mean fold increase and the seroconversion rate, quantify intra-individual increases in values. In the next sections, the analysis of these four summary statistics is explained, both for single vaccine groups and for two vaccine groups.

3.2   Geometric Mean Titres and Concentrations

Distributions of post-vaccination humoral and cellular immunogenicity values tend to be skewed to the right. Log-transformed immunogenicity values, on the other hand, usually are approximately Normally distributed. Thus, standard statistical techniques requiring Normal data can be applied to the log-transformed values. Antilogs of point and interval estimates can then be used for inference about parameters of the distribution underlying the untransformed values.

The standard statistic to summarize immunogenicity values is the geometric mean (GM), the *geometric mean titre* (GMT) if the observations are titres, or the *geometric mean concentration* (GMC) if the observations are concentrations. Let $v_1, \ldots, v_n$ be a group of n immunogenicity values. (Throughout this book, groups of observations are assumed to be independent and identically distributed (i.i.d.).) The *geometric mean* is defined as

$$GM = (v_1 \times \cdots \times v_n)^{1/n}.$$

An equivalent formula is

$$GM = \exp \sum_{i=1}^{n} (\log_e v_i / n).$$

The geometric mean response is thus on the same scale as the immunogenicity measurements.

The transformation of the immunogenicity values need not to be $\log_e$, it can be any logarithmic transformation, $\log_2$, $\log_{10}$, etc. Care should be taken that when calculating the geometric mean response the correct base is used. Thus, if $\log_{10}$ is used, the geometric mean should be computed as

$$GM = 10^{\sum_{i=1}^{n} (\log_{10} v_i / n)}.$$

If antibody titres t_i are reciprocals of twofold serial dilutions with $1:D$ as the lowest tested dilution, then a convenient log transformation is

$$u_i = \log_2[t_i/(D/2)]. \tag{3.1}$$

The u_i's are then the dilution steps: 1, 2, 3, etc. The geometric mean should be computed as

$$GM = (D/2)2^{\sum_{i=1}^{n} u_i/n}.$$

The transformation in (3.1) will be referred to as the *standard log transformation* for antibody titres.

Example 3.1. Rubella (German measles) is a disease caused by the rubella virus. In adults the disease itself is not serious, but infection of a pregnant woman by rubella can cause miscarriage, stillbirth, or damage to the foetus during the first three months of pregnancy. A haemagglutination inhibition (HI) test for rubella is often performed routinely on pregnant women. The presence of a detectable HI titre indicates previous infection and immunity to re-infection. If no antibodies can be detected, the woman is considered susceptible and is followed accordingly. Assume that in the HI test the lowest dilution is 1:8. Then the HI titres can take on the values 8, 16, 32, 64, etc. The standard log-transformed values of the titres are $\log_2(8/4) = 1, \log_2(16/4) = 2, \log_2(32/4) = 3, \log_2(64/4) = 4$, etc. The geometric mean of the five titres 8, 8, 16, 32, 64 is

$$GMT = 4 \times 2^{(1+1+2+3+4)/5}$$
$$= 18.379.$$

With the standard log transformation, differences between log-transformed values are easy to interpret: a difference of 1 means a difference of one dilution, a difference of 2 means a difference of two dilutions, etc.

A statistic often reported with the geometric mean response is the *geometric standard deviation* (GSD), which is the antilog of the sample standard deviation of the $\log_e$ transformed immunogenicity values. The statistic allows easy calculation of confidence limits for the geometric mean of the distribution underlying the immunogenicity values (the underlying geometric mean for short). Let SD be the sample standard deviation of the $\log_e$ transformed immunogenicity values, then the geometric standard deviation is

$$GSD = \exp(SD).$$

The lower and upper limit of the two-sided $100(1-\alpha)\%$ confidence interval for the underlying geometric mean e^μ are

$$LCL_{e^\mu} = GMT/GSD^{t_{n-1;1-\alpha/2}/\sqrt{n}} \tag{3.2}$$

and

$$UCL_{e^\mu} = GMT \times GSD^{t_{n-1;1-\alpha/2}/\sqrt{n}}, \tag{3.3}$$

where $t_{n-1;1-\alpha/2}$ is the $100(1-\alpha/2)$th percentile of the Student t distribution with $(n-1)$ degrees of freedom.

Example 3.1. (continued) The sample standard deviation SD of the five $\log_e$ transformed HI titres is 0.904. Thus, the geometric standard deviation is

$$GSD = e^{0.904}$$
$$= 2.469.$$

Percentiles of Student t distributions can be obtained with the SAS procedure TINV. The lower 95% confidence limit for the underlying geometric mean is

$$18.379/2.469^{2.776/\sqrt{5}} = 5.98$$

and the upper 95% confidence limit is

$$18.379 \times 2.469^{2.776/\sqrt{5}} = 56.4,$$

with $2.776 = \text{TINV}(0.975,4)$.

3.2.1 Single Vaccine Group

If the $u_i = \log_e(v_i)$ are Normally distributed with mean μ and variance σ^2, then the arithmetic mean $u.$ is a point estimate of μ, and

$$GMT = e^{u.}$$

is a point estimate of e^μ, the underlying geometric mean. The distribution of the u_i is known as the lognormal distribution.

Because the u_i are Normally distributed, confidence intervals for μ can be based on the one-sample t-test. Antilogs of the limits of the t-test based $100(1-\alpha)\%$ confidence interval for μ constitute $100(1-\alpha)\%$ confidence limits for the parameter e^μ. These confidence limits are identical to those in (3.2) and (3.3).

It should be noted that the expectation of the u_i (the mean of the lognormal distribution underlying the immunogenicity values) is not e^μ but

$$E(\mathbf{u}_i) = e^{\mu+\sigma^2}.$$

A nice property of the log-normal distribution is that e^μ is not only its geometric mean but also its median.

Example 3.2. The following data are six Th1-type IFN-γ values: 3.51, 9.24, 13.7, 35.2, 47.4 and 57.5 IU/L. The natural logarithms are 1.256, 2.224, 2.617, 3.561, 3.859 and 4.052, with arithmetic mean 2.928 and standard error 0.444. Hence,

$$GMC = e^{2.928}$$
$$= 18.7.$$

With $t_{0.975,5} = 2.571$, it follows that the two-sided 95% confidence limits for μ are

$$2.928 - 2.571(0.444) = 1.786$$

and

$$2.928 + 2.571(0.444) = 4.070.$$

Thus, the lower and upper 95% confidence limits for the geometric mean e^{μ} of the distribution underlying the IFN-γ values are

$$e^{1.786} = 5.97 \quad \text{and} \quad e^{4.070} = 58.6.$$

By definition, confidence intervals for geometric mean responses are nonsymmetrical.

3.2.2 Two Vaccine Groups

If there are two vaccine groups, statistical inference is based on the two-sample t-test, applied to the log-transformed immunogenicity values. Point and interval estimates for the difference $\Delta = \mu_1 - \mu_0$ are transformed back to point and interval estimates for the ratio $\theta = e^{\mu_1}/e^{\mu_0}$.

The standard statistic to compare two groups of immunogenicity values is the *geometric mean ratio* (GMR):

$$GMR = GM_1/GM_0,$$

where GM_1 and GM_0 are the geometric mean response of investigational and the control vaccine group, respectively. Let $u_1.$ and $u_0.$ denote the arithmetic means of the $\log_e$ transformed values of the two groups, then the following equality holds:

$$GMR = e^{u_1. - u_0.}.$$

Thus, the P-value from the two-sample t-test to test the null hypothesis $\Delta = 0$ can be used to test the null hypothesis that the underlying ratio θ equals 1.

Example 3.2. (continued) Assume that the six Th1-type IFN-γ values are to be compared with a second group of six values, and that the arithmetic mean and standard error of these ($\log_e$-transformed) values are 2.754 and 0.512, respectively. Thus,

$$GMC_0 = e^{2.754}$$
$$= 15.7$$

and the geometric mean ratio is

$$GMR = e^{2.928-2.754}$$
$$= 1.19$$
$$= 18.7/15.7.$$

The estimated standard error of the difference is 0.677. Lower and upper 95% confidence limits for the underlying geometric mean ratio are obtained as

$$e^{1.19-2.228(0.677)} = 0.73$$

and

$$e^{1.19+2.228(0.677)} = 14.9.$$

3.3 Geometric Mean Fold Increase

For some infectious diseases, pre-vaccination immunogenicity levels are not zero. An example is influenza. Recipients of influenza vaccines have usually been exposed to various influenza viruses during lifetime, by natural infections or previous vaccinations (exceptions are very young children). In that case, post-vaccination immunogenicity levels do not only express the immune responses to the vaccination but also the pre-vaccination levels. In that case, an alternative to the geometric mean titre or concentration is the geometric mean fold increase (also: mean fold increase, geometric mean fold rise).

If v_{pre} is a subject's pre-vaccination (baseline) immunogenicity value and v_{post} the post-vaccination value, then the *fold increase* is

$$fi = v_{post}/v_{pre}.$$

Fold increases express intra-individual relative increases in immunogenicity values. Just like immunogenicity values, log transformed fold increases tend to be Normally distributed, and for the statistical analysis of fold increases the methods described above for the analysis of immunogenicity values can be used. Thus, in case of a single group of fold increases the analysis will be based on the one-sample t-test, while in case of two groups of fold increases it will be based on the two-sample t-test.

3.3.1 Analysis of a Single Geometric Mean Fold Increase

The standard statistic to summarize a group of n fold increases $fi_1, \ldots, fi_n$ is the *geometric mean fold increase* (*gMFI*)

$$gMFI = \exp \sum_{j=1}^{n} (\log_e fi_j / n).$$

It is easy to show that

$$gMFI = GM_{post}/GM_{pre}.$$

Thus, the geometric mean fold increase is identical to the geometric mean of the post-vaccination values divided by the geometric mean of the pre-vaccination values. Note, though, that this equation holds only if for all n subjects both the pre- and the post-vaccination value is nonmissing. If the post-vaccination value is missing for, say, k subjects, and GM_{pre} is based on data of n subjects but GM_{post} on that of $(n-k)$ subjects, then the above equation does not hold.

Example 3.3. Consider an influenza trial in which pre- and post-vaccination anti-HA antibody levels are measured by means of the HI test. Let (5,40), (5,80), (10,160), (10,320), (20,80) and (20,640) be the pre- and post-vaccination antibody titres of the first six subjects enrolled. $GMT_{pre} = 10.0$ and $GMT_{post} = 142.5$. The fold increases are 8, 16, 16, 32, 4 and 32. The geometric mean of these six fold increases is $gMFI = 14.25$. The same value is obtained if GMT_{post} is divided by GMT_{pre}. The geometric standard deviation of the fold increases is $GSD = 2.249$. The formulae's in (3.2) and (3.3) can be used to calculate a confidence interval for the geometric mean of the distribution underlying the fold increases.

3.3.2 Analysis of Two Geometric Mean Fold Increases

To compare two groups of fold increases the two-sample t-test can be applied to the log-transformed fold increases.

In case of two groups of fold increases the following equation holds

$$\frac{gMFI_1}{gMFI_0} = \frac{GM_{post\,1}/GM_{pre\,1}}{GM_{post\,0}/GM_{pre\,0}}$$

$$= \frac{GM_{post\,1}/GM_{post\,0}}{GM_{pre\,1}/GM_{pre\,0}}$$

$$= \frac{GMR_{post}}{GMR_{pre}}.$$

Thus, the ratio of two mean fold increases (the *geometric mean fold ratio* (gMFR)) is identical to geometric mean ratio of the post-vaccination immunogenicity values divided by the geometric mean ratio of the pre-vaccination immunogenicity values. This observation has an interesting implication. Consider a randomized trial in which two vaccines are being compared. Because of the randomization, and if the sample sizes are not too small, $GM_{\text{pre } 0}$ will approximately be equal to $GM_{\text{pre } 1}$, and their ratio GMR_{pre} will be approximately be equal to 1.0. Thus, the ratio of the geometric mean fold increases will be approximately equal to the ratio of the geometric means of the post-vaccination immunogenicity values

$$gMFR = \frac{gMFI_1}{gMFI_0}$$

$$\approx GMR_{\text{post}}.$$

In other words, if there is no baseline imbalance, an analysis of the fold increases will yield a result virtually identical to that of the analysis of the post-vaccination immunogenicity values.

Example 3.3. (continued) Assume that in the trial two influenza vaccines are being compared. Let (5,40), (5,80), (10,80), (10,80) and (5,80), (5,80), (10,80), (10,160) be the pre-and post-vaccination antibody titres of the experimental and the control vaccine group, respectively. The following summary statistics are found

$$GMT_{\text{pre } 1} = GMT_{\text{pre } 0} = 7.1$$

and

$$GMT_{\text{post } 1} = 67.27 \quad \text{and} \quad GMT_{\text{post } 0} = 95.14.$$

The fold increases are: 8, 16, 8, 8 and 16, 16, 8, 16. The mean fold increases are

$$gMFI_1 = 9.51 \quad \text{and} \quad gMFI_0 = 13.45.$$

Thus, the ratio of the mean fold increases

$$gMFR = 13.45/9.51$$
$$= 1.4,$$

which is indeed identical to the ratio of the post-vaccination geometric mean titres

$$GMR_{\text{post}} = 95.14/67.27$$
$$= 1.4.$$

As explained, the reason why these two ratios are identical is that the two pre-vaccination geometric mean titres are identical.

If there is no baseline imbalance between two vaccine groups, then it will be inefficient to use the fold increases to analyze post-vaccination immunogenicity levels, because, in general, fold increases are more variable than post-vaccination immunogenicity values, resulting in larger P-values and wider confidence intervals. The explanation for this is that the log-transformed fold increase is a difference, namely between a log-transformed post-vaccination and a log-transformed pre-vaccination immunogenicity value. If the variances σ^2 at baseline and post-vaccination are the same, then the variance of the log difference is $2\sigma^2(1 - \rho)$, which implies that if the correlation ρ between the post- and the pre-vaccination values is less than 0.5, the variance of the difference is larger than that of the post-vaccination values.

3.3.3 A Misconception about Fold Increases and Baseline Imbalance

A change score is an intra-individual difference between a post- and a pre-treatment value. On a logarithmic scale, a fold increase is a change score:

$$\log fi = \log(v_{\text{post}}/v_{\text{pre}})$$
$$= \log v_{\text{post}} - \log v_{\text{pre}}.$$

If in a clinical trial the baseline values of a given characteristic (age or weight, for example) or a measurement (say, a bioassay) differ between the treatment groups, then it is said that there is baseline imbalance. Baseline imbalance matters only if the baseline value measurement is related to the primary endpoint, i.e., if it is prognostic. In that case one treatment will have a poorer prognosis than the other. The effect of the imbalance depends on the size of the imbalance and the strength of the association between the baseline value and the endpoint. If the characteristic or the baseline measurement is not prognostic, then baseline imbalance is of no concern and can be ignored.

For many bioassays, nonzero (detectable) pre-vaccination values will be predictive of the post-vaccination values because pre- and post-vaccination bioassay values tend to be positively correlated. It is often argued that imbalance in baseline values of a measurement can be dealt with by change scores, like fold increases for example. The reasoning being that if the pre-treatment values are subtracted from the post-treatment values, any bias due to baseline imbalance is eliminated. This is a fallacy. Change scores are generally correlated with pre-treatment values, the correlation often being negative. For bioassays this phenomenon is also often seen, the higher the average pre-vaccination value the smaller the average fold increase. A difference between vaccine groups in pre-vaccination state is thus predictive not only of a difference in post-vaccination state but also of a difference in fold increase, albeit in the opposite direction. Hence, in case of baseline imbalance a comparison of fold increases is in favour of the vaccine group with the smaller pre-vaccination values. Thus, it is a misconception that a fold increase analysis deals with baseline imbalance. A proper statistical technique to control for pre-vaccination state is analysis of covariance, which is discussed in Chap. 5 of this book.

3.4 Two Seroresponse Rates

3.4.1 Seroprotection Rate

For many infectious diseases, it is assumed that there is a given antibody level that is associated with protection from infection or disease. This antibody level is called the *threshold of protection*, and a subject is *seroprotected* if his antibody level is above this threshold. For influenza for example, the threshold is an anti-HA antibody level of 40, and subjects with an anti-HA antibody titre ≥ 40 are said to be seroprotected for influenza. For diphtheria and tetanus, a protection threshold found in the literature is an anti-D/anti-T antibody concentration of 0.1 IU/ml. Seroprotection is thus a binary endpoint, and the *seroprotection rate* is the proportion of vaccinated subjects who are seroprotected.

What is meant with 'associated with protection' is not always clearly defined, or understood. Sometimes it is interpreted as meaning that being seroprotected implies being fully protected against the disease. It is this interpretation that may have lead to an overestimation of the importance of the concept of seroprotection. If this interpretation ever holds can be doubted. A more reasonable interpretation is that being seroprotected means a moderate to high probability of protection. To come back to the example of influenza, the assumption is that at an anti-HA antibody titre of 40 a subject has a probability of being protected of 0.5. For a further discussion on the topic, see Sect. 3.8.

3.4.2 Seroconversion Rate

Stedman's Medical Dictionary defines *seroconversion* as development of detectable particular antibodies in the serum as a result of infection or immunization. A subject without antibodies in his serum is called *seronegative*, while a subject with antibodies is called *seropositive*. A subject's *serostatus* is his status with respect to being seropositive or seronegative for a particular antibody. A subject whose serostatus was seronegative but has become seropositive is said to have seroconverted. In an article on the safety and immunogenicity of a live attenuated human rotavirus vaccine, Vesikari and co-workers define seroconversion as appearance of serum IgA to rotavirus in post-vaccination sera at a titre of ≥ 20 U/ml in previously uninfected infants [1]. Depending on the vaccine dose, 73%–96% of the infant subjects seroconverted.

In the scientific literature, however, alternative definitions of seroconversion can be found. A popular alternative definition is: a fourfold rise (also: increase) in antibody level. This definition is often used when recipients of a vaccine may be seropositive at enrollment. The major cause of cervical cancer and cervical dysplasia (abnormal maturation of cells within tissue) is the human papillomavirus (HPV). Cervical cancer is cancer of the cervix, the lower part of the uterus. Cervical cancer develops when abnormal cells in the cervix begin to multiply abnormally. There are over 20 serotypes of HPV that affect the genital areas. Harro and co-workers, who investigated the safety and immunogenicity of a virus-like particle papillomavirus vaccine, define seropositive as an ELISA antibody titre greater than or equal to the reactivity of a standard pooled serum [2]. At study start, 6 out of the 72 females were seropositive, and seroconversion was defined as a fourfold or greater rise in titre.

Yet another definition of seroconversion is one that combines the two given above: becoming seropositive if seronegative at enrollment, or a fourfold rise if seropositive at enrollment. For example, in clinical influenza vaccine studies seroconversion is usually defined as: an anti-HA antibody titre <10 at baseline and a post-vaccination titre ≥ 40 *or* a titre >10 at baseline and at least a fourfold increase in titre post-vaccination. (In fact, the reader may note that this is a third alternative definition, because the definition thus not say: a baseline titre <10 ($=$ seronegative) and a post-vaccination titre ≥ 10 ($=$ seropositive), but, a post-vaccination titre ≥ 40 ($=$ seroprotected.))

Whichever of the above definitions is used, just like seroprotection, seroconversion is a binary endpoint, and the *seroconversion rate* – the percentage of study subjects who seroconverted – is a proportion.

3.5 Analysis of Proportions

In this section, the analysis of seroprotection and seroconversion rates is discussed. However, this will be done in the wider context of analyzing proportions. The reason for doing so is that in clinical vaccine trials binary endpoints and thus proportions are very common: proportions of subjects reporting local or systemic reactions, proportions of subjects reporting a particular adverse vaccine event, proportions of subjects remaining disease-free after vaccination, etc. Thus, the methods discussed in this section are not only applied to seroprotection and seroconversion rates but to many other kinds of rates.

3.5.1 Analysis of a Single Proportion

Null hypotheses about the rate π of a particular binary event, e.g., becoming seroprotected or having seroconverted, can be statistically tested using a test based on the binomial distribution $B(n,\pi)$, with n the number of observations. To test the null hypothesis $H_0: \pi \leq \pi_0$ against the one-sided alternative hypothesis $H_1: \pi > \pi_0$ the tail probability $\Pr(S \geq s \mid \pi_0)$ is computed, with S a $Bin(n,\pi_0)$ distributed random variable and s the observed number of events. This probability can be computed with the SAS function PROBBNML(π, n, m), which returns the probability that an observation from a BIN(n, π) distribution is less or equal to m.

Example 3.4. Feiring and co-workers report the results of a study with a meningo-coccal B vaccine in a group of 374 children, of whom 248 were randomized to the meningococcal B vaccine group and 126 to the placebo group [3]. Antibodies were measured with the serum bactericidal assay (SBA), and seroprotection was defined as a SBA titre ≥ 4. In total, 226 children received all three doses of the meningococcal B vaccine. Six weeks follow-up immunogenicity data were available for 218 children, of whom 132 were seroprotected at follow-up. Assume that the null hypothesis is that $\pi = 0.5$. The seroprotection rate is $132/218 = 0.61$, and

$$\Pr(S \geq 132|0.5) = 1 - \text{PROBBNML}(0.5, 218, 131)$$
$$= 0.0011.$$

Thus, if the null hypothesis is tested at the one-sided significance level 0.025, it can be rejected.

More usual than testing null hypotheses about a rate π is to compute a confidence interval for it.

3.5.1.1 Confidence Intervals for a Single Rate

There are numerous methods to compute confidence intervals for a single rate π. Some methods are exact, others asymptotic. There is no single superior method. Often used criteria for the evaluation of the different methods are the coverage probability and the expected width of the interval. For a comparison of seven standard methods, the reader is referred to the paper by Newcombe [4]. New methods to compute confidence intervals for single rates are still being published. In 2006, for example, Borkowf proposes a method based on adding a single imaginary failure or success [5]. Here, three methods will be discussed: the Clopper–Pearson method, the Wald method and the Wilson method.

The Clopper–Pearson method is an exact method, based on the binomial test. The lower limit of the $100(1-\alpha)\%$ Clopper–Pearson confidence interval is the largest value for π such that

$$\Pr(S \geq s|\pi) \leq \alpha/2.$$

Conversely, the upper limit of the interval is the smallest value for π such that

$$\Pr(S \leq s|\pi) \leq \alpha/2.$$

The FREQ procedure of SAS returns Clopper–Pearson confidence limits if requested (use the option `bin` with the `tables` statement). Furthermore, formula for the limits exists. If $F_{k,l;1-\alpha/2}$ is the $100(1-\alpha/2)$th percentile of the F distribution with k numerator degrees of freedom and l denominator degrees of freedom, then the Clopper–Pearson lower confidence limit is

$$LCL_{CP} = \frac{s}{s + (n - s + 1)f_L},$$

with $f_L = F_{2(n-s+1),2s;1-\alpha/2}$. The Clopper–Pearson upper confidence limit is

$$UCL_{CP} = \frac{(s+1)f_U}{(n-s)+(s+1)f_U},$$

with $f_U = F_{2(s+1),2(n-s);1-\alpha/2}$.

Example 3.4. (continued) In the study 132 out of 218 children were seroprotected. Upper percentiles of F distributions can be computed with the SAS function FINV. The upper percentile is

$$\begin{aligned} f_L &= \text{FINV}(0.975, 2 \times (218 - 132 + 1), 2 \times 132) \\ &= \text{FINV}(0.975, 174, 264) \\ &= 1.307. \end{aligned}$$

So that

$$\begin{aligned} LCL_{CP} &= \frac{132}{132 + (218 - 132 + 1)1.307} \\ &= 0.537. \end{aligned}$$

Similarly, the upper percentile is

$$\begin{aligned} f_U &= \text{FINV}(0.975, 2 \times (132 + 1), 2 \times (218 - 132)) \\ &= \text{FINV}(0.975, 266, 172) \\ &= 1.318 \end{aligned}$$

and

$$\begin{aligned} UCL_{CP} &= \frac{(132 + 1)1.318}{(218 - 132) + (132 + 1)1.318} \\ &= 0.671. \end{aligned}$$

Thus, the two-sided 95% Clopper–Pearson interval for the probability of being seroprotected is $(0.537, 0.671)$.

A drawback of the Clopper–Pearson method is that it conservative in the sense that the coverage probability of the interval is *at least* the nominal value, $1 - \alpha$. To overcome this disadvantage, mid-P confidence limits have been proposed [6,7]. The lower limit of the $100(1-\alpha)\%$ mid-P confidence interval is the largest value for π such that

$$\frac{\Pr(S = s|\pi)}{2} + \Pr(S > s|\pi) \leq \alpha/2.$$

The upper is the largest value for π such that

$$\frac{\Pr(\mathbf{S} = s|\pi)}{2} + \Pr(\mathbf{S} < s|\pi) \leq \alpha/2.$$

The coverage probability of the mid-P confidence interval is, on average, equal to the nominal value but depending on n and π may be much larger or smaller. For this reason, the mid-P confidence method is not recommended (but for a different view refer [4]).

The Wald and the Wilson method are both asymptotic methods. The Wald method is the simpler of the two:

$$LCL_{\text{Wald}}, UCL_{\text{Wald}} = r \pm z_{1-\alpha/2} SE(r),$$

with $r = s/n$, the observed success rate and

$$SE(r) = \sqrt{r(1-r)/n}$$

its estimated standard error. The coverage probability of the Wald interval is on average too low and may be very low if π is in the vicinity of zero or one. When the continuity correction is used the coverage probability is improved but for extreme π it may be far of the nominal value. In vaccine development, this is a serious drawback because seroprotection rates, for example, are often close to one, while adverse vaccine event rates can be close to zero.

An asymptotic method with an average coverage probability close to the nominal value is the Wilson method. Under the null hypothesis H_0: $\pi = \pi_0$ the statistic

$$Z = \frac{r - \pi_0}{SE_0(r)}$$

is approximately standard Normally distributed, with

$$SE_0(r) = \sqrt{\pi_0(1-\pi_0)/n}$$

the standard error of r under the null hypothesis. According to one of the first principles of statistics, the range of all values for π_0 that are not rejected at the significance level α constitute a $100(1-\alpha)\%$ confidence interval for π. To find the limits of this range, the following equation must be solved

$$\frac{(r - \pi_0)^2}{\pi_0(1-\pi_0)/n} = z_{1-\alpha/2}^2.$$

This leads to the following limits

$$LCL_{\text{Wilson}}, UCL_{\text{Wilson}} = \frac{-B \pm \sqrt{B^2 - 4AC}}{2A},$$

with

$$A = n + z_{1-\alpha/2}^2, \quad B = -(2s + z_{1-\alpha/2}^2), \quad C = s^2/n.$$

Example 3.4. (continued) With $z_{0.975} = 1.96$, it follows that

$$A = (218 + 1.96^2)$$
$$= 221.84$$

$$B = -(2 \times 132 + 1.96^2)$$
$$= -267.84$$

$$C = 132^2/218$$
$$= 79.93.$$

Thus, asymptotic two-sided 95% confidence limits for π are

$$LCL_{\text{Wilson}} = \frac{267.84 - \sqrt{267.84^2 - 4 \times 221.84 \times 79.93}}{2 \times 221.84}$$
$$= 0.539$$

and

$$UCL_{\text{Wilson}} = \frac{267.84 + \sqrt{267.84^2 - 4 \times 221.84 \times 79.93}}{2 \times 221.84}$$
$$= 0.668.$$

For the example, the confidence limits based on the Wilson method are almost identical to the limits based on the Clopper–Pearson method. These Wilson-type confidence limits can be requested in the SAS procedure FREQ by using the option `bin (wilson)` with the `tables` statement.

3.5.2 Comparing Two Proportions

There are two statistics to compare two proportions r_1 and r_0, e.g., two seroprotection rates or two seroconversion rates. These statistics are the rate difference

$$RD = r_1 - r_0$$

and the rate ratio

$$RR = r_1/r_0.$$

The rate difference is an estimator of the risk difference

$$\Delta = \pi_1 - \pi_0$$

and the rate ratio is an estimator of the relative risk

$$\theta = \pi_1/\pi_0.$$

To test the null hypothesis that $\Delta = 0$ is equivalent to testing that $\theta = 1$. Both null hypotheses can be tested with Pearson's chi-square test or, in case of small sample sizes, an exact test, either the well-known Fisher's exact test, or, preferably, the less well-known Suissa and Shuster test. This latter test is discussed in Sect. 3.5.3.

To compute confidence intervals for Δ or θ, both asymptotic and exact methods are available. For both parameters, two types of asymptotic intervals that can be found in the statistical and epidemiological literature are the familiar Wald-type intervals and the less familiar Wilson-type intervals. The Wilson-type intervals should be the intervals of choice because their coverage is superior to that of the Wald-type intervals.

Asymptotic confidence limits for the risk difference Δ are often computed as

$$RD \pm z_{1-\alpha/2} SE(RD), \tag{3.4}$$

where

$$SE(RD) = \sqrt{r_1(1-r_1)/n_1 + r_0(1-r_0)/n_0}, \tag{3.5}$$

with n_1 and n_0 the sizes of the two groups. These Wald-type confidence limits are available in the SAS procedure FREQ (use the option `riskdiff` with the `tables` statement). The Wald approach may mean disagreement between Pearson's chi-square statistic to test the null hypothesis $H_0: \Delta = 0$ and the confidence limits. This is because the test statistic is computed under the null hypothesis, with the standard error of RD also estimated under the null hypothesis:

$$SE_0(RD) = \sqrt{r(1-r)(1/n_1 + 1/n_0)}, \tag{3.6}$$

where

$$r = \frac{s_1 + s_0}{n_1 + n_0}$$

and s_1 and s_0 the observed numbers of events. Then

$$\chi^2_{\text{Pearson}} = \frac{RD^2}{SE_0^2(RD)}.$$

The Pearson statistic and the Wald-type confidence interval are thus based on different estimates of the standard error of the rate difference – $SE_0(RD)$ versus $SE(RD)$ – which explains the occasional disagreement between the test statistic and the interval. Another drawback of the Wald approach is that $SE(RD)$ cannot be calculated if either r_1 or r_0 equals zero or one.

An approach that does not suffer from these drawbacks was proposed by Mietti-nen and Nurminen [8]. In their approach, based on the Wilson method, the limits of the two-sided $100(1-\alpha)\%$ confidence interval for Δ are those values Δ_0 that satisfy the equation

$$\frac{RD - \Delta_0}{SE_{\Delta_0}(RD)} = \pm z_{1-\alpha/2}, \tag{3.7}$$

where

$$SE_{\Delta_0}(RD) = \sqrt{\tilde{R}_1(1 - \tilde{R}_1)/n_1 + \tilde{R}_0(1 - \tilde{R}_0)/n_0}. \tag{3.8}$$

$\tilde{R}_1$ and $\tilde{R}_0$ are constrained maximum likelihood estimates of π_1 and π_0, with as constraint

$$\tilde{R}_1 - \tilde{R}_0 = \Delta_0.$$

Miettinen and Nurminen give a closed-formed solution for $\tilde{R}_0$, which is reproduced in Appendix B of this book. The confidence limits have to be found iteratively. A simple iterative approach is the following. If 95% confidence limits are required, with a precision of, say, three decimals, then, to find the upper limit of the interval, evaluate the test statistic on the left-hand side of (3.7) for $\Delta_0 = RD + 0.001$, $\Delta_0 = RD + 0.002$, etc., until the test statistic exceeds 1.96. The upper limit is the largest tested value for Δ for which the test statistic is less than 1.96. To find the lower limit, evaluate the test statistic for $\Delta_0 = RD - 0.001$, $\Delta_0 = RD - 0.002$, etc., until the test statistic falls below -1.96. The lower limit is the smallest value for Δ for which the test statistic is greater than -1.96.

Example 3.5. Consider a randomized immunogenicity trial in which both seated-tection rates are equal to 1.0, say, $r_1 = 48/48$ and $r_0 = 52/52$. The Wald approach does not allow calculation of a 95% confidence interval for Δ, but the Wilson approach does: $(-0.074, 0.068)$.

The standard approach to compute an asymptotic confidence interval for the rel-ative risk θ is to compute Wald-type confidence limits based on the log-transformed rate ratio RR, which are then back transformed, the so-called logit limits. In this approach, the standard error of $\log_e RR$ is estimated as

$$SE(\log_e RR) = \sqrt{1/s_1 - 1/n_1 + 1/s_0 - 1/n_0}. \tag{3.9}$$

The logit limits of the two-sided $100(1-\alpha)\%$ Wald-type confidence interval for θ are

$$\exp[\log_e RR \pm z_{1-\alpha/2} SE(\log_e RR)]. \tag{3.10}$$

These Wald-type confidence limits for θ are also available in the SAS procedure FREQ (use the option `relrisk` the `tables` statement).

For the rate ratio also, Miettinen and Nurminen derived a Wilson-type confidence interval. The approach is the same as for the rate difference. The limits of the two-sided $100(1 - \alpha)\%$ Wilson-type confidence interval for θ are the values θ_0 that satisfy the equation

$$\frac{r_1 - \theta_0 r_0}{SE_{\theta_0}(r_1 - \theta_0 r_0)} = \pm z_{1-\alpha/2},$$

with

$$SE_{\theta_0}(r_1 - \theta_0 r_0) = \sqrt{\tilde{R}_1(1 - \tilde{R}_1)/n_1 + \theta_0^2 \tilde{R}_0(1 - \tilde{R}_0)/n_0}. \tag{3.11}$$

Again, $\tilde{R}_1$ and $\tilde{R}_0$ are constrained MLEs of π_1 and π_0, with as constraint

$$\tilde{R}_1 = \theta_0 \tilde{R}_0.$$

For the closed-form solution for $\tilde{R}_0$, see Appendix B.

Example 3.5. (continued) Because $SE(\log_e RR) = 0.0$, the Wald approach does not allow calculation of a confidence interval for θ. The two-sided 95% Wilson-type confidence interval for θ is (0.926, 1.073).

The Wilson-type confidence interval for θ cannot be evaluated when either s_1 or s_0 equals zero.

As a final remark, it is important to be aware that the estimator RR is biased, that it overestimates θ. The explanation is that it is nonlinear in the maximum likelihood estimator r_0. The bias will be nonnegligible when the control rate θ_0 approaches zero and n_0 is small to intermediate [9]. This may be the case in vaccine field efficacy trials (see Chap. 7), with often very low attack rates, and in safety analyzes (see Chap. 9), when comparing adverse events rates between vaccine groups. At least two bias corrections exists [9]. The simplest but very effective bias correction for the rate ratio is to add one imaginary event to the control group, i.e., to set s_0 to $(s_0 + 1)$ and n_0 to $(n_0 + 1)$. This correction is known as Jewell's correction. Simulation results on the performance of this correction are given in Appendix C. Somewhat surprisingly perhaps, Jewell's correction does not improve the performance of the Wilson-type confidence interval for the risk ratio, in the sense that it gives better coverage. In fact, the uncorrected confidence procedure provides near nominal coverage while the corrected procedures could give subnominal coverage.

3.5.3 The Suissa and Shuster Exact Test for Comparing Two Proportions

Fisher's test is an exact test for comparing two proportions, based on conditioning on the margins. An alternative to this test is a test based on the maximization method, the Suissa and Shuster test, which is also an exact test [10]. This test is conditional on the sample sizes being fixed, but it does not condition on the number of observed cases. The Suissa and Shuster test has been shown to be more powerful than Fisher's exact test [11]. The test has an attractive property, which Fisher's test lacks, namely that it can be used to test null hypotheses of the form $H_0: \Delta = \Delta_0$, with $\Delta_0 \neq 0.0$, and $H_0: \theta = \theta_0$, with $\theta_0 \neq 1.0$. This means that the test can be used to derive exact confidence intervals for the risk difference and the relative risk.

To compute the P-value for the Suissa and Shuster exact test, the procedure proposed by Chan is given [12]. *Step 1* is that the appropriate Z statistic to compare two numbers of events is identified. To test the null hypothesis $H_0: \Delta = \Delta_0$ the Z statistic is

$$Z = \frac{RD - \Delta_0}{SE_{\Delta_0}(RD)},\tag{3.12}$$

where $SE_{\Delta_0}(RD)$ is the standard error (3.8). When the null hypothesis is $H_0: \theta = \theta_0$, the Z statistic is

$$Z = \frac{r_1 - \theta_0 r_0}{SE_{\theta_0}(r_1 - \theta_0 r_0)},\tag{3.13}$$

with $SE_{\theta_0}(r_1 - \theta_0 r0)$ the standard error (3.11). For $\Delta_0 = 0.0$ and $\theta_0 = 1.0$, both statistics reduce to the standard Z statistic to compare two proportions:

$$Z = \frac{RD}{SE_0(RD)},\tag{3.14}$$

with $SE_0(RD)$ the standard error (3.6). Let Z_{obs} be the value for the appropriate Z statistic for the observed numbers of events (s_1, s_0).

Step 2 is that for every possible 2×2 table the value Z_{ij} for the Z statistic is computed, and that all combinations (i, j) of numbers of events with $|Z_{ij}| \geq |Z_{obs}|$ are identified.

Step 3 is to find a P-value for testing the null hypothesis. The two-sided P-value for the Suissa and Shuster test is defined as

$$\text{P-value} = \max_{\{\pi_0 \in D\}} \Pr(|Z| \geq |Z_{obs}||\pi_0).$$

For a given value for π_0, $\Pr(|Z| \geq |Zobs||\pi_0)$ is the sum of the probabilities of those 2×2 tables with $|Z_{ij}| \geq |Z_{obs}|$. Under the null hypothesis, these probabilities are the products of two binomial probabilities:

$$\binom{n_1}{i} \pi_1^i (1 - \pi_1)^{n_1 - i} \times \binom{n_0}{j} \pi_0^j (1 - \pi_0)^{n_0 - j}$$

with $\pi_1 = (\pi_0 + \Delta_0)$ for the Z statistic in (3.8), $\pi_1 = \theta_0 \pi_0$ for the Z statistic (3.11) and $\pi_1 = \pi_0$ for the Z statistic (3.6).

The parameter π_0 is a nuisance parameter. The domain D for π_0 is the continuous interval $[0, 1-\Delta_0]$ if $\Delta_0 > 0$ and $[\Delta_0, 1]$ if $\Delta_0 < 0$; if $\theta_0 > 1$ then D is the interval $[0, 1/\theta_0]$, and if $\Delta_0 < 1$ then D is $[0,1]$; if $\Delta_0 = 0$ and $\theta_0 = 1$ then D is the interval $[0,1]$. Chan proposes to divide the domain for π_0 in a large number of equally spaced intervals and calculate the probability at every increment, e.g., $(0.001, 0.002, \ldots, 0.999)$ if the domain is $[0,1]$, an approach which would provide

sufficient accuracy for most practical uses. The computer must be instructed to set
the Z statistic (3.8) to zero for the 2×2 table with numbers of events (0,0), and to set
the Z statistic (3.11) to zero for 2×2 tables with numbers of events (0,0) or (n_1, n_0).

If the null hypothesis $H_0: \Delta \leq \Delta_0$ or $H_0: \theta \leq \theta_0$ is to be tested against the
one-sided alternative $H_1: \Delta > \Delta_0$ or $H_1: \theta > \theta_0$, then the one-sided P-value is

$$\text{P-value} = \max_{\{\pi_0 \in D\}} \Pr(Z \geq Z_{\text{obs}} | \pi_0).$$

The one-sided P-value to test the null hypothesis $H_0: \Delta \geq \Delta_0$ or $H_0: \theta \geq \theta_0$ is to
be tested against the one-sided alternative $H_1: \Delta < \Delta_0$ or $H_1: \theta < \theta_0$ is

$$\text{P-value} = \max_{\{\pi_0 \in D\}} \Pr(Z \leq Z_{\text{obs}} | \pi_0).$$

Lydersen, Fagerland and Laake compared the performance of Fisher's exact test and
the Suissa and Shuster and other similar tests [11]. The performance of the Suissa
and Shuster exact test is superior to that of Fisher's exact test, which is conservative.
The performance of the mid-P version of Fisher's exact test comes close to that of
the Suissa and Shuster test. They advise that Fisher's exact test should no longer be
used.

Example 3.6. Chan cites a challenge study on the protective efficacy of a recombinant protein influenza vaccine. In the study 15 vaccinated and 15 placebo subjects
were challenged with a weakened A-H1N1 influenza virus strain. After 9 days the
observed rates of any clinical illness were 7/15 in the vaccine group and 12/15 in
the control group, the placebo group. For this data, Fisher's exact test yields a one-
sided P-value of 0.064 and a two-sided P-value of 0.128. For the null hypothesis
$H_0 : \Delta = 0.0$, the observed value for the Z statistic is $Z_{\text{obs}} = -1.894$. The Suissa
and Shuster test yields a one-sided P-value of 0.034 and a two-sided P-value of
0.068. The asymptotic P-value for Z_{obs} is the one for Pearson's chi-square statistic,
which for the example data equals 0.058.

Exact confidence intervals for the risk difference or the relative risk can be obtained
by testing the appropriate null hypothesis against the one-sided alternative for
subsequent values for Δ_0 or θ_0.

Example 3.6. (continued) For the challenge data, the exact two-sided 95% confidence interval for the relative risk θ is quickly found once a SAS code to compute
the Suissa and Shuster exact P-values has been written. The one sided P-value for
the null hypothesis $H_0: \theta \leq 0.260$ is 0.0231, and that for $H_0: \theta \leq 0.261$ is
0.0263. Thus, the lower limit of the exact confidence interval is 0.261. The one-sided
P-values for the null hypotheses $H_0: \theta \geq 1.037$ and $H_0: \theta \geq 1.038$ are 0.02503
and 0.0248, respectively. Thus, the upper limit of the exact confidence interval is
1.037. For comparison, the asymptotic 95% Wilson-type confidence interval for θ
is (0.300, 1.019).

3.6 Multiple Co-Primary Endpoints and the Intersection–Union Test

In clinical vaccine trials, it is not uncommon that there are multiple co-primary endpoints. As an example, consider a trial with an experimental combination vaccine containing different serotypes of the same organism, e.g., a pneumococcal vaccine. If the aim of the trial is to compare the experimental vaccine to the control vaccine for all serotypes, then the number of co-primary endpoints will be equal to the number of serotypes. Another example is a trial with the aim to compare an experimental vaccine to a control vaccine for both the seroprotection rate and the geometric mean response, in which case there are two co-primary endpoints.

When there is more than one primary endpoint, the multiplicity issue must be addressed. Here, one particular type of multiplicity will be considered, namely the scenario that the objective of the trial is to demonstrate that the experimental vaccine is superior to the control vaccine simultaneously for all co-primary endpoints. For this scenario, a much applied approach is the one based on the intersection–union (IU) test [13]. In this approach, for each of the k co-primary endpoints, the component null hypothesis is tested at the significance level α, and superiority of the investigational vaccine to the control vaccine is claimed only if all k component null hypotheses are rejected. On the plus side of the IU test is its simplicity, which makes the test easy to explain to nonstatisticians. On the negative side is that the test is known to be conservative. Only when the k endpoints are perfectly correlated the level of the test will be exactly α, in all other cases the level will be less than α. When the k endpoints are independent the level of the test will be as low as α^k.

Several alternatives to the IU test have been proposed. For the case that the data follow a multivariate Normal distribution, for example, Laska and colleagues show that under mild conditions a test known as the *min* test is uniformly the most powerful test [14]. The *min* test statistic $Z_{\min}$ is defined as

$$Z_{\min} = \min\{Z_1, \ldots, Z_k\},$$

with Z_i a test statistic to test the ith component null hypothesis. The sampling distribution of the test statistic is, however, complicated and depends amongst others on the covariance structure of the endpoints, which limits the applicability of the test.

For two interesting discussions on multiple co-primary endpoints and statistical power, the reader is referred to the papers by Chuang-Stein and colleagues and Senn and Bretz [15, 16].

3.7 The Reverse Cumulative Distribution Plot

The *reverse cumulative distribution plot* is a graphic tool to display the distribution of immunogenicity values. It is particularly useful for visual comparisons of distributions between vaccine groups. The plot became quickly popular after a lucid presentation of its properties by Reed, Meade and Steinhoff [17].

In Fig. 3.1, four examples of a reverse cumulative distribution (RCD) curve are shown. The x-axis represents the immunogenicity values, and the scale of the axis is usually logarithmic. The y-axis represents the percentage of subjects having at least that immunogenicity value. Thus, to the value x on the x-axis corresponds the percentage of subjects having a log-transformed immunogenicity value greater or equal to x. By definition the curve begins at 100%, and then descends down, from left to right. The lowest point on the curve is the percentage of subjects having a log-transformed immunogenicity value equal to the highest observed value. The plot is called the reverse cumulative distribution plot because it reverses the cumulative distribution. The median log-transformed immunogenicity level is the value on the x-axis below the y-axis value of 50%.

If the distribution of the log transformed immunogenicity values is symmetric with little variability around the mean, then the middle section of the RCD curve will be steep (curve A). If, on the other hand, the variability is large then the middle section of the curve will be less steep (curve B). In the extreme, if the distribution of the log-transformed immunogenicity values is more or less uniform, then the curve will be approximately a downward-sloping straight line. If the distribution of the log-transformed immunogenicity values is skewed to the right, with a large fraction of the subjects having a value near the high end, then the curve will be rectangular, i.e., remaining high and flat with a rapid descend near the end (curve C). If the distribution of the log transformed immunogenicity values is skewed to the left, thus if there are many subjects with a low value, then the curve will be similar to curve D.

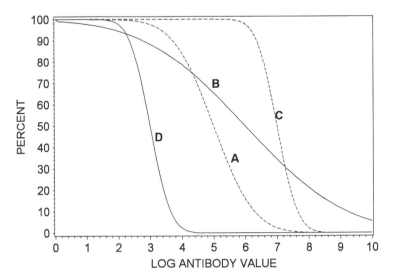

Fig. 3.1 Four examples of a RCD curve

If the RCD curve for one vaccine is above the curve for another vaccine, and the two curves do not intersect (say, A and D in Fig. 3.1), then every percentile of distribution of the immunogenicity values of vaccine A is higher than the corresponding percentile for vaccine D, meaning that vaccine A induced the higher immune responses. If the two curves intersect, then one vaccine induced both more lower and higher immune responses. If (the y-value of) the point of intersection is above 50% (as for B and A in Fig. 3.1) than this is in favour of vaccine B, because compared to group A, in group B there would be a larger fraction of subjects with a high immune response, while a point of intersection below 50% would be in favour of vaccine A.

3.8 Discussion

When it has been demonstrated that an experimental vaccine is superior to a control vaccine with respect to, say, mean antibody response, it is often assumed that the conclusion can be generalized to protection, i.e., that the experimental vaccine is also superior to the control with respect to protection against disease or infection. Nauta, Beyer and Osterhaus investigated this for the case of influenza vaccines [18]. With the help of a simple statistical model, they show that the relationship between antibody level and protection from influenza is more complicated than perhaps envisioned. Their model predicts that the relationship depends not only on the mean but also on the standard deviation of the log-transformed antibody values. Until their publication, this dependency had been largely overlooked. It is this dependency that complicates the interpretation. They observe, for example, that if the mean antibody level of both the experimental and the control group are high, a positive difference in mean level implies a positive difference in the fraction protected subjects, unless the difference in mean level is small to moderate in combination with a large, positive difference in standard deviation. Interpretation of differences in fractions seroprotected subjects is even more challenging. Their model predicts that differences in the fraction seroprotected subjects cannot be interpreted without taking into account the mean antibody levels and standard deviations. This sheds a whole new light on the usefulness of the concept of seroprotection.

It is not unreasonable to assume that these observations also hold for many other vaccines. The implication would be that the standard methods discussed in this chapter are in need of improvement in the sense that they compare means rather than (simultaneously) means and standard deviations of distributions. An early attempt in this direction was made by Lachenbruch, Rida and Kou, who outline a statistical method based on measuring the similarity between the scales and shapes of antibody level distributions [19]. Computationally, their method is complex. For example, critical values for the test statistic have to be simulated. Nevertheless, their approach – evaluating simultaneously the means and the standard deviations of immunogenicity distributions – is certainly worth to be further developed.

3.9 Sample Size Estimation

3.9.1 Comparing Two Geometric Mean Responses

There are many different formulae for sample size estimation for parallel group trials with Normal data. The simplest but least accurate one and which can be found in any basic text on statistics, is the well-known one based on a Normal approximation. The most accurate one, involving no approximations, is based on a noncentral t distribution, see Formula (3.8) in the book by Julious [20]. Sample sizes based on this formula can be estimated with procedure POWER of SAS. Two parameters have to be specified, the difference Δ of the log-transformed underlying geometric means and the within-groups standard deviation σ of the log-transformed immunogenicity values. The value for σ may be taken from previous trials. Alternatively, the following formula can be used to obtain a, conservative, value for σ:

$$4\sigma \approx [\log(\text{largest expected immunogenicity value}) \ minus$$
$$\log(\text{lowest expected immunogenicity value})].$$

This formula is based on the fact that in case of Normal data approximately 95% of the observations will fall in the range

$$\mu - 2\sigma \text{ to } \mu + 2\sigma.$$

To obtain a less conservative value, in the formula above the multiplier 4 should be substituted with 6.

Example 3.7. Consider a clinical trial in which two influenza vaccines are to be compared, a licensed one and a new, investigational vaccine, with the antibody response as measured by the HI test as primary endpoint. Assume that the investigator believes that the new vaccine is considerably more immunogenic than the licensed one, and that he expects geometric mean ratio to be ≥ 2.0. In influenza vaccine trials HI titres $>5,120$ are rare, and the lowest possible value is usually 5, i.e., half of the reciprocal of the starting dilution, 10. Thus

$$\Delta = \log_e 2.0 = 0.69,$$

and a conservative value for σ is

$$\sigma = \frac{\log_e 5,120 - log_e 5}{4} = 1.73.$$

Below the SAS code to estimate the number of subjects per vaccine group for a statistical power of 0.9 is given. The required sample size is found to be 134 subjects per group, 268 subjects in total.

SAS Code 3.1 *Sample Size Calculation for Comparing two Geometric Mean Responses*

```
proc power;
   meandiff=0.69 stddev=1.73
   power=0.9 npergroup=.;
run;
```

SAS Output 3.1

```
              The POWER Procedure
        Two-sample t Test for Mean Difference
              Fixed Scenario Elements

    Distribution                    Normal
    Method                          Exact
    Mean Difference                 0.69
    Standard Deviation              1.73
    Nominal Power                   0.9
    Number of Sides                 2
    Null Difference                 0
    Alpha                           0.05

        Computed N Per Group

        Actual      N Per
        Power       Group
        0.902       134
```

For the reverse approach, finding the statistical power for a given sample size, in the SAS code the variable `power` should be set to missing (.) and the variable `npergroup` to the proposed number of subjects per group.

3.9.2 Comparing Two Proportions

For the estimation of the sample size required to compare two proportions also, numerous formulae exist. All these formulae are asymptotic, and a detailed discussion of them can be found in Chap. 4 of the book by Fleiss, Levin and Paik [21]. The formulae they advise, (4.14) for equal sample sizes and (4.19) for unequal sample sizes, have been included in the POWER procedure. For a discussion on exact power calculations for 2×2 tables, the reader is referred to the paper by Hirji and colleagues [22].

Example 3.7. (continued) Assume that the investigator wants to compare seroprotection rates rather than geometric mean titres, and that he expects an increase in the probability of seroprotection from 0.85 to 0.95. Below the SAS code to calculate the number of subjects per vaccine group for a statistical power of 0.9 is given. The required sample size is found to be 188 subjects per group.

SAS Code 3.2 *Sample Size Calculation for Comparing two Proportions*

```
proc power;
    twosamplefreq test=pchi
    groupproportions=(0.85,0.95)
    power=0.9 npergroup=.;
run;
```

SAS Output 3.2

```
                    The POWER Procedure
        Pearson Chi-square Test for Two Proportions
                    Fixed Scenario Elements

        Distribution            Asymptotic normal
        Method                  Normal approximation
        Alpha                   0.05
        Group 1 Proportion      0.85
        Group 2 Proportion      0.95
        Nominal Power           0.9
        Number of Sides         2
        Null Proportion Difference 0

            Computed N Per Group

            Actual      N Per
            Power       Group

            0.901        188
```

3.9.3 Sample Size Estimation for Trials with Multiple Co-Primary Endpoints

Estimation of the power of a trial with multiple co-primary endpoints is a nontrivial problem. The key of the problem is that the power is critically dependent on the correlation between the endpoints. Different assumptions about the correlation can lead to substantially different sample size estimates. The power will be the highest when the endpoints are strongly correlated but will be low when the endpoints are minimally correlated. In practice, the correlations between the endpoints will often be unknown. If the objective of the trial is to demonstrate statistical significance for all k co-primary endpoints (see Sect. 3.6), then an often applied approach is to obtain a lower bound for the global power P using the following inequality

$$P \geq \prod_{i=1}^{k} P_i, \tag{3.15}$$

with P_i the power of the trial for the ith endpoint. This inequality requires the assumption that all endpoints are nonnegatively correlated.

An inequality that does not require any assumptions about the correlations is

$$P \geq \sum_{i=1}^{k} P_i - (k-1). \tag{3.16}$$

(See Appendix D for a proof of this inequality.) When k is large, both inequalities require that the P_i's must be close to 1.0 to be secured of a lower bound exceeding 0.8. Assume that the global power should be at least 0.9. This requirement will be met if

$$P_i \geq (0.9)^{1/k} \quad i = 1, \ldots, k.$$

If $k = 5$, then the P_i should be at least 0.979.

Example 3.8. Consider a comparative trial with two co-primary endpoints, the geometric mean concentration and the seroprotection rate. Assume that a sample size of 2×150 subjects is being considered, and that with the help of the SAS codes in the previous sections the power for the first co-primary endpoint has been estimated as 0.93 and for the second co-primary endpoint as 0.90. Under the (reasonable) assumption that the two endpoints are nonnegatively correlated, a lower bound for the global statistical power is

$$P \geq 0.93 \times 0.90$$
$$= 0.837.$$

If inequality (3.16) is used instead, the lower bound for the global power is

$$0.93 + 0.90 - 1 = 0.830.$$

When data of previous clinical trials are available, an alternative method to estimate the global power is Monte-Carlo simulation. As a simple example, consider an open study with the aim to demonstrate the immunogenicity of a new formulation of a particular vaccine, by showing that both the seroprotection and the seroconversion rate exceed pre-defined bounds. A large number of studies – minimally 5,000 – is simulated. Per simulated study a random draw with replacement of size n from the database is made, and the result of the study is considered significant if the observed seroprotection and seroconversion rate exceed the pre-defined bounds. An estimate of the global power of the design for sample size n is the fraction of simulated studies with a significant result.

Chapter 4
Antibody Titres and Two Types of Bias

4.1 Standard Antibody Titres versus Mid-Value Titres

Statisticians have pointed out that when antibody titres are determined using the standard definition (see Sect. 2.1.2), the reciprocal of the highest dilution at which the assay read-out did occur, the true titre value is underestimated [23]. This is easy to see. By definition, the true antibody titre τ lies between the standard titre t_s and the reciprocal of the next dilution, r_n:

$$t_s \leq \tau < r_n.$$

Thus, for most serum samples the standard antibody titre will be lower than the true titre. This means that if standard antibody titres are used, the geometric mean titre will underestimate the geometric mean of the distribution underlying the antibody values.

There are antibody assays that try to correct for this bias. An example is the interpolated serum bactericidal assay (SBA) to demonstrate humoral immune responses induced by meningococcal vaccines. Meningococcal disease is caused by the bacterium *Neisseria meningitidis*, also known as *meningococcus*. Attack rates of the disease are the highest among infants aged younger than two years and adolescents between 11 and 19 years of age. The disease can cause substantial mortality. Five serogroups, A, B, C, Y and W135, are responsible for virtually all cases of the disease. The standard SBA titre is defined as the reciprocal of the highest dilution of serum immediately preceding the 50% survival/kill value for colony-forming units (50% cut-off). The interpolated SBA titre is calculated using a formula that calculates the percentage kill in dilutions on either side of the 50% cut-off. The titre is the reciprocal of the dilution of serum at the point where the antibody curve intersects the 50% cut-off line.

Another approach to correct for the bias is changing the definition of the antibody titre to

$$\text{antibody titre} = \sqrt{t_s r_n},$$

the geometric mean of the standard antibody titre and the reciprocal of the next dilution. For titres based on serial two-fold dilutions r_n equals $2t_s$, in which case

J. Nauta, *Statistics in Clinical Vaccine Trials*, DOI 10.1007/978-3-642-14691-6_4,
© Springer-Verlag Berlin Heidelberg 2011

the alternative definition of the antibody titre becomes

$$\text{antibody titre} = \sqrt{2}t_s.$$

For reasons explained below, Nauta and De Bruijn propose to call this the *mid-value definition of antibody titres* [24]. Mid-value antibody titres are higher than standard titres. For example, if the predefined dilutions are 1:4, 1:8, 1:16, etc., and the standard antibody titre for a serum sample is 64, then the mid-value titre for the sample is the geometric mean of 64 and 128

$$\sqrt{64 \times 128} = 90.5$$
$$= \sqrt{2} \times 64.$$

On a logarithmic scale, the mid-value antibody titre t_{mv} is the mid-point between t_s and r_n:

$$\log t_{mv} = (\log t_s + \log r_n)/2.$$

Hence the name: mid-value antibody titre. Nauta and De Bruijn were not the first to promote mid-value antibody titres (although they coined the term), an early suggestion to use mid-value titres rather than standard titres can be found in a paper by Lyng and Weis Bentzonyn [25].

In almost all practical situations, the mid-value definition reduces the bias of standard antibody titres, meaning that on average the mid-value titre is closer to the true titre than the standard titre, i.e.,

$$|t_{mv} - \tau| \le |t_s - \tau|.$$

A sufficient condition for the mid-value definition to reduce the bias is

1. The log-transformed antibody titres are Normally distributed, and
2. The dilutions are predefined, and
3. The distance between two consecutive log-transformed dilution factors is small compared to the range of observed log-transformed titre values

For post-vaccination titres, this is usually the case. (For pre-vaccination titres, often condition 1 or 3 is not met.)

In case of serial two-fold dilutions, when the antibody titres are determined using the standard definition and the primary outcome measure is the geometric mean titre, there is no need to calculate the individual mid-value titres. Let GMT_s be the geometric mean of a single group of standard antibody titres. Then the geometric mean of the corresponding mid-value titres is

$$GMT_{mv} = \sqrt{2}GMT_s.$$

A similar expression holds for the confidence limits for GMT_{mv}.

For the geometric mean fold increase, which can be expressed as the ratio of a post- and a pre-vaccination geometric mean titres (see Sect. 3.3), the bias correction is also not needed, because

$$gMFI_{mv} = \frac{GMT_{mv\ post}}{GMT_{mv\ pre}}$$

$$= \frac{\sqrt{2}GMT_{s\ post}}{\sqrt{2}GMT_{s\ pre}}$$

$$= \frac{GMT_{s\ post}}{GMT_{s\ pre}}$$

$$= gMFI_s.$$

When the summary statistic of interest is the geometric mean fold increase, both definition, the standard and the mid-value one, will produce the same result. The same holds true for the geometric mean ratio, the ratio of two independent geometric mean titres:

$$GMR_{mv} = \frac{GMT_{mv\ 1}}{GMT_{mv\ 0}}$$

$$= \frac{\sqrt{2}GMT_{s\ 1}}{\sqrt{2}GMT_{s\ 0}}$$

$$= \frac{GMT_{s\ 1}}{GMT_{s\ 0}}$$

$$= GMR_s.$$

Here also both definitions produce the same result.

4.2 Censored Antibody Titres and Maximum Likelihood Estimation

If the number of dilutions in an antibody assay is limited, then it may happen that at the highest tested dilution the assay read-out did not occur. In that case, often the titre is set to the reciprocal of the highest dilution. The result of this practice is bias. The geometric mean titre will be underestimated, and the assumption of Normality for the distribution of the log-transformed titres will no longer hold.

Example 4.1. In Fig. 4.1, the histogram of the frequency distribution of log transformed post-vaccination measles HI antibody titres of a hypothetical group of 300 children is displayed. The starting dilution was 1:4, and as log transformation the standard transformation – $\log_2(titre/2)$ – was used. The arithmetic mean of the log transformed titres is 5.80, with estimated standard deviation 2.49. This arithmetic mean corresponds to a geometric mean titre of

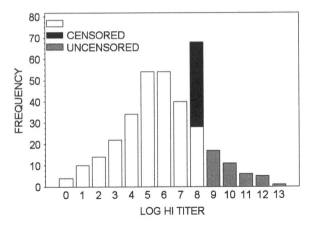

Fig. 4.1 Uncensored and censored frequency distribution of log transformed measles HI antibody titres

$$2 \times 2^{5.80} = 111.4.$$

The median and the maximum of the titres are 128 and 16,384, respectively. Next, suppose that the highest tested dilution would have been 1:512 rather than a much higher dilution, and that titres above 512 would have been set to 512. The result of this censoring is also shown in Fig. 4.1. The bars above 9, 10, 11, 12 and 13 are added to the bar above 8. Due to the censoring the frequency distribution is no longer symmetrical and thus not longer Normally shaped. The arithmetic mean of the censored log-transformed titres is 5.53 ($GMT = 92.4$), with estimated standard deviation 2.05. Both estimates are smaller than the estimates based on the uncensored data.

In the following sections, it will be explained how this bias due to censoring can be eliminated [26].

4.2.1 ML Estimation for Censored Normal Data

A censored observation is an observation for which a lower or an upper limit is known but not the exact value. An example of a censored observation is a value below the detection limit of a laboratory test. The upper limit of the test result is known, the detection limit, but not the test result itself. If for an observation only an upper limit is known, the observation is called left censored. Observation for which only a lower limit is known are called right censored. Censored observations are often assigned the value of the limit. If this is not taken into account in the statistical analysis, estimates will be biased. A powerful statistical method to eliminate this bias is maximum likelihood (ML) estimation for censored data. As an introduction to ML estimation for censored antibody titres, in this section ML estimation for censored Normal data is explained.

Let $x_1, \ldots, x_n$ be a group of noncensored continuous $N(\mu, \sigma^2)$ distributed observations. The log likelihood function is

$$LL(\mu, \sigma) = \sum_{i=1}^{n} \log f(x_i; \mu, \sigma),$$

where $f(x; \mu, \sigma)$ is the Normal density function. The ML estimates of μ and σ are those values that maximize the log likelihood function. For Normal data, the ML estimates are the arithmetic mean and the estimated standard deviation (but with the $(n-1)$ in the denominator replaced by n.)

Next, suppose that r of the observations are right-censored. Let $x_1, \ldots, x_{n-r}$ be the noncensored observations and $x_{n-r+1}, \ldots, x_n$ the right-censored observations. Then the log likelihood function becomes

$$LL(\mu, \sigma) = \sum_{i=1}^{n-r} \log f(x_i; \mu, \sigma) + \sum_{i=n-r+1}^{n} \log[1 - F(x_i; \mu, \sigma)],$$

where $F(x; \mu, \sigma)$ is the Normal distribution function. Thus, for a censored observation x, the density for x is replaced with the probability of an observation beyond x. Again, the ML estimates of μ and σ are those values that maximize the log likelihood function, and they are unbiased estimates of these parameters.

Finally, suppose that there are l left-censored observations: $x_1, \ldots, x_l$. The log likelihood function for Normal data with both left- and right-censored observations is

$$LL(\mu, \sigma) = \sum_{i=1}^{l} \log F(x_i; \mu, \sigma) + \sum_{i=l+1}^{n-r} \log f(x_i; \mu, \sigma)$$

$$+ \sum_{i=n-r+1}^{n} \log[1 - F(x_i; \mu, \sigma)].$$

Maximum likelihood estimation for censored Normal data can be intuitively understood as follows. For a series of values for μ and σ a Normal curve is fitted to the histogram of the frequency distribution of the observation, and it is checked how well the curve fits to the data. This includes a comparison of the areas under the left and the right tail of the fitted curve with the areas of the histogram bars below or above the censored values. The censored tails are reconstructed, and correct estimates of the mean and standard deviation of the distribution are obtained. The ML estimates of μ and σ are those values that give the best fit, and they are found by iteration.

Maximum likelihood estimation for censored observations was first introduced for the analysis of survival data, where it is used to deal with censored survival times [27].

4.2.2 ML Estimation for Censored Antibody Titres

To compute ML estimates for censored, log-transformed antibody titres, the SAS procedure LIFEREG can be used. This is a procedure to fit probability distributions to data sets with censored observations. A wide variety of probability distributions can be fitted, including the Normal distribution.

Before the log likelihood function for censored observations can be applied to log-transformed antibody titres, a further modification is needed. In the previous section, it was assumed that the data were censored continuous Normal observations. Log transformed antibody titres, however, are not continuous observations, they are so-called interval censored observations. An interval censored observation is an observation for which the lower and the upper value is known but not the exact value. If this is not taken into account, i.e., if the values are treated as if continuous, the ML estimates procedure LIFEREG returns will be invalid.

As explained in Sect. 4.1, the true titre value τ lies between the standard titre t_s and the reciprocal r_n of the next dilution:

$$t_s \leq \tau < r_n.$$

Thus, let t_i be an interval censored standard titre, with r_i the reciprocal of the next dilution. The second term of the log likelihood function, the term for the noncensored observations becomes

$$\sum_{i=l+1}^{n-r} \log \left[F(\log r_i; \mu, \sigma) - F(\log t_i; \mu, \sigma) \right].$$

The first term of the log likelihood function, the term for the left-censored observations, becomes

$$\sum_{i=1}^{l} \log F(\log r_L; \mu, \sigma),$$

where r_L is the reciprocal of the starting dilation. The third term of the log likelihood function, the term for the right-censored observations, becomes

$$\sum_{i=n-r+1}^{n} \log \left[1 - F(\log r_H; \mu, \sigma) \right],$$

where r_H is the reciprocal of the highest dilution tested.

In procedure LIFEREG this modification can be handled by the `model` statement with the `lower` and `upper` syntax; `lower` and `upper` are two variables containing the lower and the upper ranges for the observations. For an interval censored standard titre `lower` $= \log t_i$ and `upper` $= \log r_i$; for left-censored standard titres `lower` has to be set to missing (interpreted by the procedure LIFEREG as minus infinity) and `upper` to $\log r_L$; for right-censored standard titres lower $= \log r_H$ and upper has to be set to missing (interpreted as plus infinity.) By definition, all observations below the detection limit are left-censored.

Example 4.1. (continued) The starting dilution was 1:4, and thus antibody titres less or equal to 4 are to be considered as left censored. To visualize this, in Fig. 4.1 the bar above 0 has to be added to the bar above 1. Below a SAS code to fit a Normal distribution to the censored log transformed antibody titres in Fig. 4.1 is given.

SAS Code 4.1 *Fitting a Normal Distribution to the Censored Antibody titres of Fig. 4.1*

```
data;
  input titre count;
  logtitre=log(titre/2)/log(2);/*standard log transformation*/
  do i=1 to count;
    if (titre=4)   then lower=.; else lower=logtitre;
    if (titre=512) then upper=.; else upper=logtitre+1;
  output;
  end;
  datalines;
  4    14
  8    14
  16   22
  32   34
  64   54
  128  54
  256  40
  512  68
  run;

proc lifereg;
    model (lower,upper)= / d=normal;
  run;
```

SAS Output 4.1A

```
                    Analysis of Parameter Estimates


                         Standard  95% Confidence    Chi-
    Parameter DF Estimate  Error      Limits      Square Pr > ChiSq

    Intercept 1   6.2270  0.1443  5.9443  6.5098 1863.11   <.0001
    Scale     1   2.3918  0.1242  2.1604  2.6480
```

Two parameters are estimated, an intercept, which is the ML estimate of μ, and a scale parameter, which is the ML estimate of σ. For both parameters, two-sided 95% confidence limits are given.

The reader will have observed that the μ estimated above is the mean of the distribution underlying the log-transformed mid-value titres (Sect. 4.1), and not the mean of the distribution underlying the log transformed standard titres. This is due to the values assigned to the SAS variables lower and upper, which are consistent with the mid-value definition. To estimate the μ consistent with the definition of standard titres, in the SAS code above lower has to be set to logtitre-0.5 and upper to logtitre+0.5. If done so, the following output is obtained

SAS Output 4.1B

```
              Analysis of Parameter Estimates

                      Standard  95% Confidence   Chi-
    Parameter DF Estimate  Error     Limits   Square Pr > ChiSq

    Intercept 1  5.7270   0.1443  5.4443  6.0098 1575.93  <.0001
    Scale     1  2.3918   0.1242  2.1604  2.6480
```

The ML estimate of μ is now consistent with the standard titre definition. (This value could have of course also be obtained by subtracting 0.5 from the ML estimates in SAS Output 4.1A: $6.2270 - 0.5 = 5.7270$.) Note that the correction does not have an effect on the ML estimate of σ.

The ML estimates 5.73 and 2.39 are in good agreement with the estimates based on the uncensored data, 5.80 and 2.49. This demonstrates the powerful tool ML estimation for censored observations is.

Above as log transformation the standard transformation $\log t = \log_2 [t/(D/2)]$ was used, with D the starting dilution factor. A general SAS code to fit a Normal distribution to the censored $\log_e$ transformed serial two-fold antibody titres is presented below.

SAS Code 4.2 *Fitting a Normal Distribution to Censored Serial Two-fold Antibody titres*

```
data;
   input titre;
   midvalue=1;  /* 1 for mid-value definition, 0 otherwise */
   rsd=4;       /* reciprocal starting dilution */
   rhd=512;     /* reciprocal highest dilution */
   if (midvalue) then
   do;
       if (titre=rsd) then lower=.; else lower=log(titre);

       end;
     else
   do;
         if (titre=rsd) then lower=.;
```

```
            else lower=(log(titre)+log(titre/2))/2;
            if (titre=rhd) then upper=.;
            else upper=(log(titre)+log(2*titre))/2;
        end;
datalines;
.
.
.
run;

proc lifereg;
    model (lower,upper)= / d=normal;
run;
```

The approach discussed above can be readily extended to the case of two vaccine groups. Let `group` be the SAS variable for the groups, taking the value 1 for the experimental group and 0 for the control group. Then the procedure LIFEREG statement in SAS code 4.1 should be changed to

```
proc lifereg;
    model (lower,upper)=group /d=normal;
run;
```

Chapter 5
Adjusting for Imbalance in Pre-Vaccination State

5.1 Imbalance in Pre-Vaccination State

For some infectious diseases, pre-vaccination antibody levels are not zero. Not all vaccines offer life-long protection, and a number of diseases require re-vaccinations throughout life. Antibody levels prior to re-vaccination with, for example, a tetanus, a diphtheria, a pertussis or a tick borne encephalitis vaccine will often not be zero. If pre-vaccination (baseline) antibody levels are not zero, then the post-vaccination values do not only express the immune responses to the vaccination, but also the subjects' pre-vaccination state. This can complicate the interpretation of a difference in post-vaccination antibody values between vaccine groups. If there is imbalance in pre-vaccination state, i.e., if there is a difference in baseline antibody values between groups, then part of the post-vaccination difference can be explained by the pre-vaccination difference. In case of a positive imbalance in pre-vaccination state, the post-vaccination difference may overestimate the immunogenicity of an investigational vaccine.

A popular approach to correct for imbalance in pre-vaccination state is analysing the fold increases instead of the post-vaccination antibody values. In Sect. 3.3.3 it was argued that the reasoning behind this approach – if pre-treatment values are subtracted from post-treatment values, any bias due to baseline imbalance is eliminated – is incorrect. If post- and pre-vaccination antibody values are positively correlated, and they usually are, then a positive baseline difference is predictive of a positive post-vaccination difference, but it is also predictive of a negative difference in mean fold increase. Thus, an analysis of fold increases does not solve the problem of imbalance in pre-vaccination state.

Consider the scatterplot in Fig. 5.1. Shown are log-transformed post-vaccination antibody values (y-axis) versus log-transformed pre-vaccination values (x-axis) for a given infectious disease. All points fall above or on the diagonal line of equality, because as a rule a post-vaccination antibody value will be larger than or equal to the pre-vaccination value. Because biologically there is a maximum to the amount of antibodies that the body can produce, all points fall below a horizontal line, the asymptote. The vertical distance between a point and the diagonal line is the log-transformed fold increase. The larger the log-transformed pre-vaccination antibody

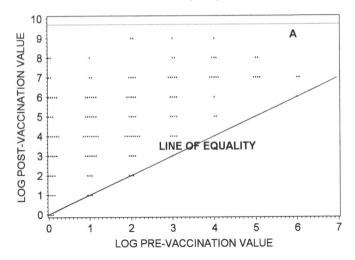

Fig. 5.1 Post-vaccination antibody values versus pre-vaccination values

value, the smaller, on average, the log-transformed fold increase. Figure 5.1 is thus a graphical illustration of the negative correlation between pre-vaccination antibody values and fold increases.

Before proceeding to discuss two statistical techniques to adjust for baseline imbalance, the question needs to be addressed if such an adjustment is required. Strictly speaking, the answer is no. Statistical theory does not require baseline balance. In a randomized trial, baseline balance is not a peremptory requirement to yield valid results. Even with randomization, treatment groups will never be fully balanced with respect to all prognostic factors. If randomization is applied, groups will be equal on average, i.e., over all possible randomizations. Random between-group outcome differences are allowed for in the statistics computed from the data, and the inference drawn from them is correct, on average.

Nevertheless, there is a general consensus that the credibility of a statistical analysis is increased if baseline imbalance for known prognostic factors is corrected for, as an improvement over just relaying on the randomization.

5.2   Adjusting for Baseline Imbalance

There are two major statistical techniques to adjust for baseline imbalance: stratification and analysis of covariance.

In a stratified trial, subjects with a similar baseline value are assigned to the same stratum. If subjects are randomized per stratum, then treatment groups will be comparable with respect to the variable used for stratification. In clinical vaccine trials, however, pre-randomization stratification by baseline antibody value is rarely applied. One reason for this may be that this approach requires two instead of one baseline visit. During the first visit, a blood sample for baseline antibody titration is drawn. Baseline samples are then sent to a laboratory for antibody determination, which may take several weeks. And then, when the assay results have been received by the site, the subjects must return for a second baseline visit. If the vaccine contains different serotypes of the same organism, then pre-randomization stratification is also very difficult. An alternative to pre-randomization stratification is post-randomization stratification, with the strata being defined after the trial has been completed, during the statistical analysis of the data. A much applied approach is to divide the baseline values into the four quartiles, which then serve as strata. In both cases, pre- or post-vaccination stratification, a stratified analysis is performed and the baseline imbalance is eliminated.

The second technique to adjust for baseline imbalance is analysis of covariance, sometimes referred to as regression control. (Analysis of covariance can be viewed as a limiting case of post-randomization stratification, with the observed baseline values as the strata.) Two parallel lines are fitted to the outcome data, one for the investigational treatment group and one for the control group, with regression on the baseline data. The treatment effect is then estimated by the vertical distance between the fitted lines. If there is baseline imbalance, a comparison of the outcome means will be affected by the difference in mean baseline value. Affected in the sense that the difference of the outcome means will overestimate the treatment effect. A comparison between the two groups for subjects with the same baseline value would be the solution. This is what analysis of covariance does.

5.3 Analysis of Covariance for Antibody Values

The simplest case of analysis of covariance for antibody values is the analysis of data from a single vaccine study. A linear regression model is fitted to the data with the log-transformed post-vaccination antibody value (y) as the response (also: outcome, dependent) variable and the log-transformed pre-vaccination antibody value (x) as the predictor (also: independent) variable:

$$\mathbf{Y} = \beta_0 + \beta_1 x + \mathbf{E}_x. \tag{5.1}$$

The intercept β_0 and the slope β_1 are regression parameters to be estimated from the data. If the observations are antibody titres and the standard log transformation is used, then the intercept β_0 is the mean of the distribution underlying the y's of the seronegative subjects, i.e., the subjects with a pre-vaccination antibody titre of $(D/2)$, with D the starting dilution factor. Then the slope β_1 is the average increase in the y's per dilution step.

Note that

$$(D/2)2^{\beta_0 + \beta_1 x}$$

is the geometric mean of the distribution underlying the antibody values of subjects with a log-transformed pre-vaccination value of x. Thus, the regression parameters can be used for inference about the untransformed antibody values.

5.3.1 A Solution to the Problem of Heteroscedasticity

The residual $\mathbf{E}_x$ is the difference between the actual y and the value predicted by the fitted model. The usual assumption is that $\mathbf{E}_x$ is Normally distributed about a mean value of 0 with variance σ_x^2. Another usual assumption is that of homoscedasticity. This is the assumption that σ_x^2 does not depend on x, but that $\sigma_x^2 = \sigma^2$. However, Fig. 5.1 shows that in case of log-transformed antibody values this assumption may not hold, but that σ_x^2 decreases with increasing x. A dependency of σ_x^2 on x is called heteroscedasticity. If this heteroscedasticity is ignored, i.e., if the regression model is fitted under the assumption of homoscedasticity, then the resulting parameter estimates, confidence intervals and P-values will be invalid. But a solution to this problem exists once it is realized that the heteroscedasticity can be modelled.

Let A be the upper limit for the log-transformed post-vaccination antibody values, i.e., the horizontal asymptote in Fig. 5.1. For Normal data, a crude formula for the range of the values to be observed is

$$\text{range} \approx c \times \text{standard deviation},$$

with the constant c often being set to either 4 or 6. The range for the y's for subjects with a log-transformed pre-vaccination value of x is $(A\text{-}x)$, and thus

$$(A - x) \approx c\sigma_x,$$

or, after taking squares on both sides of the equation

$$A^2 - 2Ax + x^2 \approx c^2\sigma_x^2.$$

This can be rewritten as

$$\sigma_x^2 \approx \sigma^2(1 + c_1 x + c_2 x^2), \tag{5.2}$$

where

$$\sigma^2 = (A/c)^2, c_1 = -2/A, c_2 = 1/A^2.$$

Equation (5.2) thus gives a model for the variance of $\mathbf{E}_x$.

5.3.2 Fitting the Variance Model for Heteroscedasticity

The variance model in (5.2) cannot be fitted with SAS, but an almost similar model can. This model is

$$\sigma_x^2 = \sigma^2 \exp(C_1 x + C_2 x^2). \tag{5.3}$$

To see that these two models are very similar, consider the case that

$$A = 10, c_1 = -2/10 = -0.2, c_2 = 1/10^2 = 0.01.$$

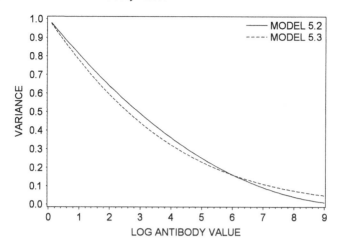

Fig. 5.2 Graphical comparison of variance models (5.2) and (5.3)

This model is plotted in Fig. 5.2, where for convenience σ has been set to 1. Also plotted in Fig. 5.2 is the model in (5.3) with

$$C_1 = -0.24, C_2 = -0.011.$$

The two curves almost coincide.

The model in (5.3) has the nice property that it cannot be negative or zero, which is a guarantee for more stable variance estimates. Below a SAS code to fit this variance model is given.

SAS Code 5.1 *ANCOVA for Log-Transformed Antibody Values, under the Assumption of Heteroscedasticity*

```
proc mixed;
   model y=x / solution;
   repeated / local=exp(x x2);
run;
```

When the above code is run, estimates of $\beta_0, \beta_1, \sigma, C_1$ and C_2 are returned. Because C_1 and C_2 are nuisance parameters, their estimates are of no special interest.

The regression model in (5.1) was fitted to the data of Fig. 5.1, both under the assumption of homoscedasticity and under the assumption that for the σ_x^2 the variance model in (5.3) applied. In Table 5.1, parameter estimates are given. As expected, the estimate of σ under the assumption of homoscedasticity is smaller than the estimate under the assumption of heteroscedasticity. If the standard log transformation is used, the intercept β_0 is the mean of the distribution underlying the y's of the seronegative subjects. This parameter is often of special interest

Table 5.1 Regression parameter estimates for the data in Fig. 5.1

	n	β_0	β_1	σ
Assuming Homoscedasticity	163	3.715	0.669	1.867
Assuming Heteroscedasticity	163	3.765	0.642	2.151
Seronegative Subjects only	55	3.818		2.212

because the seronegative subjects are the ones most in need of improved immunity. Alternative estimates of β_0 and σ are therefore the sample mean and standard deviation of the y's of the seronegative subjects, which are also given in Table 5.1. That this latter standard deviation is an estimate not only of σ_0 but also of σ is easy to see, for

$$\sigma_0^2 = \sigma^2 \exp(C_1 0 + C_2 0^2)$$
$$= \sigma^2.$$

The estimator for σ_0 based on the regression approach is the more precise of the two. This can be seen by comparing the two standard errors. The standard error based on the y's of the seronegative subjects is

$$2.212/\sqrt{55} = 0.298.$$

In contrast, the standard error of the regression estimate of σ_0 – taken from the SAS output (not shown) – is 0.218. The explanation for this difference is that the first standard error is based on only 55 observations, while the second one is based on all 165 observations.

5.3.3 ANCOVA for Comparative Clinical Vaccine Trials

Suppose that in a randomized clinical vaccine trial the pre-vaccination geometric mean titres are 13.2 for the investigational vaccine group and 7.9 for the control vaccine group, expressing a moderate baseline imbalance in favour of the investigational vaccine.

The post-vaccination geometric mean titres are 286.3 and 112.8, respectively. The uncorrected post-vaccination geometric mean ratio thus is

$$\text{uncorrected } GMR = 286.3/112.8$$
$$= 2.54.$$

To correct for this baseline imbalance, the following regression model was fitted to the data

$$\mathbf{Y} = \beta_0 + \beta_1 group + \beta_2 x + \mathbf{E}_x, \tag{5.4}$$

with $group = 1$ if the subject was vaccinated with the investigational vaccine and $group = 0$ if the subject was vaccinated with the control vaccine. Because parallel regression lines are assumed, β_1 is the expected difference between the two groups for subjects with the same baseline titre. The baseline-corrected geometric mean ratio is

$$\text{baseline-corrected } GMR = 2^{\beta_1}.$$

The baseline-corrected geometric mean ratio should be smaller than the uncorrected one.

The model in (5.4) was fitted under the assumption of heteroscedasticity, using the following SAS code.

SAS Code 5.2 *ANCOVA for Log-Transformed Antibody Values assuming Parallel Regression Lines*

```
proc mixed;
    model y=group x / solution;
    repeated / local=exp(x x2);
run;
```

The following output is produced:

SAS Output 5.2

Effect	Estimate	Standard Error	DF	t Value	Pr >|t|
Intercept	4.1538	0.2148	243	19.34	<.0001
group	0.7931	0.2769	243	2.86	0.0045
x	0.6079	0.07202	243	8.44	<.0001

The estimated baseline-corrected geometric mean ratio is

$$2^{0.7931} = 1.73,$$

a value, which is in indeed smaller than the uncorrected ratio, 2.54.

Above the assumption of parallel regression lines was made. If this assumption is in doubt, it can be checked by adding the interaction term group*x to the model.

SAS Code 5.3A *ANCOVA for Log-Transformed Antibody Values assuming Non-Parallel Regression Lines*

```
proc mixed;
    model y=group x group*x / solution;
    repeated / local=exp(x x2);
run;
```

SAS Output 5.3A

Effect	Estimate	Standard Error	DF	t Value	Pr >|t|
Intercept	3.9608	0.2326	242	17.03	<.0001
group	1.1792	0.3306	242	3.57	0.0004
x	0.8217	0.1214	242	6.77	<.0001
group*x	-0.3157	0.1473	242	-2.14	0.0331

The interaction term is significant. For subjects vaccinated with the investigational vaccine, the regression line is somewhat flatter than that for subjects vaccinated with the control vaccine.

When lines are not parallel, the distance between the lines, and thus the baseline-corrected geometric mean ratio, becomes dependent on x. When the interaction model is fitted to the data, β_1 is the expected difference between the two groups for seronegative subjects. For the example data, the estimate for β_1 and thus that for the geometric mean ratio

$$2^{1.1792} = 2.26$$

is statistically significant. It may, however, be that the group of special interest is not the seronegative subjects but, say, the subjects with as baseline value the threshold of protection T_P. In that case, the regression model to be fitted becomes

$$Y = \beta_0 + \beta_1 group + \beta_2 z + \beta_3 group \times z + E_x. \tag{5.5}$$

where $z = (x - \log T_P)$. This model can be fitted with the following SAS code

SAS Code 5.3B *ANCOVA for Log-Transformed Antibody Values assuming Non-Parallel Regression Lines*

```
proc mixed;
    model y=group z group*z / solution;
    repeated / local=exp(x x2);
run;
```

Note that here also the variance of the error term is defined as a function of x, not of z.

SAS Output 5.3B

```
                       Standard
Effect      Estimate    Error      DF     t Value    Pr >|t|

Intercept    6.4260    0.3219     242      19.96      <.0001
group        0.2320    0.3761     242       0.62      0.5379
x            0.8217    0.1214     242       6.77      <.0001
group*x     -0.3157    0.1473     242      -2.14      0.0331
```

The only two estimates that change are those for β_0 and β_1. For subjects with the baseline value equal to the threshold of protection, the geometric mean ratio

$$2^{0.2320} = 1.17$$

is not statistically significant.

Chapter 6
Vaccine Equivalence and Noninferiority Immunogenicity Trials

6.1 Equivalence and Noninferiority

So far, it has been silently assumed that the objective of the trial was to demonstrate that one vaccine is superior to another with respect to the induced immunogenicity. Trials with this objective are called superiority trials. In an equivalence trial, the objective is not to demonstrate that two vaccines are different but that they are more or less similar, while in a noninferiority trial the objective is to demonstrate that one vaccine is not less immunogenic than the other.

Wang and colleagues indentify four types of equivalence and noninferiority designs for vaccine immunogenicity trials: vaccine bridging trials, combination vaccine trials, vaccine concomitant use trials and vaccine lot consistency trials [28]. In a *vaccine bridging trial*, two formulations of a vaccine are being compared. The reason for the study could be a change in the manufacturing process, a change in the vaccine formulation, or a change in storage conditions of the vaccine. An example of the first is when a second manufacturing site is opened, and it has to be demonstrated that the immunogenicity of the formulation produced at the new site is comparable to that of the formulation produced at the old site. An example of the second reason is the removal of a constituent from the formulation. Until recently many vaccines contained thiomersal as preservative. After the thiomersal controversy (see Sect. 9.1), manufacturers were requested to remove the preservative from their childhood vaccines and today all childhood vaccines are thiomersal-free.

In a *combination vaccine trial*, immune responses are compared between a combined vaccine and the separate but simultaneously administered monovalent vaccines. A combination vaccine is usually intended to reduce the number of injections required. For example, in 2005 the Food and Drug Administration (FDA) approved the licensure of a combination MMRV (measles, mumps, rubella, varicella) vaccine for children aged twelve months through twelve years, as alternative for two separate MMR and V vaccines. (However, soon after licensure the manufacturer of the combination vaccine withdrew the vaccine because of a safety issue, see Sect. 9.1.) Before a combination vaccine is licensed it has to be demonstrated that the combination is not less immunogenic or less safe than the monovalent vaccine.

J. Nauta, *Statistics in Clinical Vaccine Trials*, DOI 10.1007/978-3-642-14691-6_6, © Springer-Verlag Berlin Heidelberg 2011

67

In the past, combination of whole live vaccines resulted in a reduced immune response due to immunological inference between vaccine viruses.

A *vaccine concomitant use trial* is used to compare the concomitant administration of two or more vaccines and the separate administration of the vaccines. The intention is usually to see if the number of vaccination visits can be reduced.

A vaccine manufacturer who wants to license his vaccine must demonstrate that the manufacturing process is stabile, i.e., that consistent lots can be produced. This has to be demonstrated by both analytical and clinical testing. The clinical testing is done in a so-called *vaccine lot consistency trial*, and the objective of such a study is to show that the lots are similar with respect to the induced immunogenicity.

The first three types of trials are usually designed as noninferiority studies, while vaccine lot consistency trials are an example of an equivalence study. But, because the concept equivalence was introduced before that of noninferiority, first equivalence studies are discussed.

6.2 Equivalence and Noninferiority Testing

6.2.1 Basic Concepts

Let $\Delta = \mu_1 - \mu_0$ be the difference between (the expected means of) two treatments. In a superiority trial, the null hypothesis is $H_0 : \Delta = 0$, which is usually tested against the two-sided alternative $H_1 : \Delta \neq 0$, or, less common, against a one-sided alternative, say, $H_1 : \Delta > 0$. In contrast, in an equivalence trial the objective is not to demonstrate that two treatments are different but that they are more or less similar, meaning that $\Delta \approx 0$. More or less similar, because to proof exact equality would be impossible. To be more precise, the objective of an equivalence trial is to demonstrate that, on average, two treatments differ no more than by a fixed amount δ, the equivalence margin. The two treatments are considered equivalent if $|\Delta| < \delta$. To demonstrate equivalence of two treatments Schuirmann's two one-sided tests (TOST) procedure is often used [29]. There are two null hypotheses associated with the procedure, namely that the difference between the two treatments exceeds the equivalence margin: $H_{01}: \Delta \geq \delta$ and $H_{02}: \Delta \leq -\delta$. These null hypotheses are tested against the alternatives $H_{11}: \Delta < \delta$ and $H_{12}: \Delta > -\delta$. To demonstrate equivalence both null hypotheses have to be rejected. If both H_{01} and H_{02} are tested at the significance level α, this approach corresponds to checking that the two-sided $100(1-2\alpha)\%$ confidence interval for Δ lies within the equivalence range $-\delta$ to $+\delta$.

A somewhat controversial issue in equivalence trials is the choice of the significance level α, with the question being whether the level should be set to 0.05 or 0.025. Schuirmann himself suggested $\alpha = 0.05$, and in pharmacokinetics studies for example, this has become the standard. In non-phase I clinical studies, however, the preference seems to be $\alpha = 0.025$. In a superiority trial in which an active treatment is compared with a placebo, most regulatory agencies will require a two-sided significance level of 0.05. Given that the only outcome of interest is where the active treatment is significantly better than the placebo, the risk for the regulatory agency is at most 0.025. In an equivalence trial, the risk for the agency is that the true null hypothesis (either H_{01} or H_{02}) is falsely rejected. If both null hypotheses are tested at significance level, then this risk is at most α. And for this reason, it is often argued that in equivalence trails the significance level should be set to 0.025, for the sake of consistency. To stress this, in this chapter equivalence (and noninferiority) will be tested at the significance level $\alpha/2$.

Noninferiority trials are a special case of equivalence trials, the one-sided version, so to speak. To objective of a noninferiority trial is to demonstrate that one treatment is not less than another by more than a small amount, $-\delta$, the noninferiority margin. The null hypothesis of a noninferiority trial is that $H_0: \Delta \leq -\delta$, which is tested against the alternative $H_1: \Delta > -\delta$. If the null hypothesis is tested at the significance level $\alpha/2$, then this procedure leads to the same conclusion as checking that the lower bound of the one-sided $100(1-\alpha/2)\%$ confidence interval for Δ falls to the right of $-\delta$.

6.2.2 Equivalence and Noninferiority Testing for Normal Data

An alternative way to explain testing for equivalence is that the null hypothesis $H_0 : |\Delta| \geq \delta$ is tested against the alternative $H_1 : |\Delta| < \delta$. If both groups of n_1 and n_0 observations are Normally distributed, this null hypothesis can be tested using the following test statistic:

$$Z_{EQ} = \frac{\delta - |D_{10}|}{SE(D_{10})}, \qquad (6.1)$$

where $D_{10} = x_{1.} - x_{0.}$, the difference between the arithmetic means of the two groups of observations, and $SE(D_{10})$ the standard error of the difference:

$$SE(D_{10}) = \sqrt{SD_1^2/n_1 + SD_0^2/n_0},$$

with SD_1 and SD_0 the sample standard deviations of the two groups. If $|D_{10}| > \delta$, then the data are in favour of the null hypothesis, in which case Z_{EQ} will be negative. If, on the other hand, $|D_{10}| < \delta$ then Z_{EQ} will be positive, and the smaller $|D_{10}|$ the larger Z_{EQ}. The null hypothesis is thus rejected for large positive values of the test statistic, which can be compared with the $100(1 - \alpha/2)$th percentile of the standard Normal distribution. If equal variances are assumed, then an alternative estimator for the standard error is

$$SE_P(D_{10}) = SD\sqrt{1/n_1 + /n_0}, \qquad (6.2)$$

where SD is the pooled sample standard deviation. If this standard error is used, then the value for the test statistic can be compared with the $100(1 - \alpha/2)$th percentile of the t distribution with $(n_1 + n_0 - 2)$ degrees of freedom.

In a noninferiority trial, the null hypothesis $H_0: \Delta \leq -\delta$ is tested against the alternative $H_1: \Delta > -\delta$. This null hypothesis can be tested using the following test statistic:

$$Z_{NI} = \frac{D_{10} + \delta}{SE(D_{10})}. \tag{6.3}$$

The data are in favour of the null hypothesis if $D_{10} \leq -\delta$, i.e., if $D_{10} + \delta \leq 0$, and then Z_{NI} will be negative. Z_{NI} will be positive if the data are in favour of the alternative hypothesis, i.e., if $D_{10} > -\delta$. Thus here also, the null hypothesis is rejected for large positive values of the test statistic. Values for Z_{NI} can be compared either with $100(1 - \alpha/2)$th percentile of the standard Normal distribution, or the $100(1 - \alpha/2)$th percentile of the t distribution with $(n_1 + n_0 - 2)$ degrees of freedom, if the standard error (6.2) is used.

6.2.3 The Confidence Interval Approach to Equivalence and Noninferiority Testing

If $H_0 : |\Delta| \geq \delta$ is tested at the significance level $\alpha/2$, then the null hypothesis that the two treatments are not equivalent is rejected if $Z_{EQ} > z_{1-\alpha/2}$. This will be the case if and only if

$$-\delta < D_{10} \pm z_{1-\alpha/2}SE(D_{10}) < \delta.$$

Thus, testing for equivalence involves checking that the two-sided $100(1 - \alpha)\%$ confidence interval for Δ falls in the equivalence range $-\delta$ to $+\delta$. It is this confidence interval approach that is most often used in scientific publications rather than the hypothesis testing approach. If equal variances are assumed then the two-sided $100(1 - \alpha)\%$ confidence interval can be based on the t distribution, in which case it should be checked that

$$-\delta < D_{10} \pm t_{1-\alpha/2;n_1+n_0-2}SE_P(D_{10}) < \delta.$$

The null hypothesis that one treatment is less than another can be tested by checking that the lower bound of the one-sided $100(1 - \alpha/2)\%$ confidence interval for Δ falls to the right of $-\delta$:

$$-\delta < D_{10} - z_{1-\alpha/2}SE(D_{10})$$

or

$$-\delta < D_{10} - t_{1-\alpha/2;n_1+n_0-2}SE_P(D_{10}).$$

6.3 Equivalence and Noninferiority Vaccine Trials with a Geometric Mean Response as Outcome

If the outcome measure of a randomized clinical vaccine trial is a geometric mean response – a geometric mean titre or geometric mean concentration –, then the parameter of interest is the geometric mean ratio θ, the ratio of the geometric means e^{μ_1} and e^{μ_0} of the distributions underlying the immunogenicity values. Requiring that the confidence interval for the ratio θ falls within a pre-specified range is the same as requiring that the confidence interval for the difference Δ of the arithmetic means μ_1 and μ_0 of the distributions underlying the $\log_e$-transformed values falls within the log-transformed range. This means that equivalence can be stated both in terms of θ and Δ: if (λ_1, λ_2) is an equivalence range for θ, then $(\log_e \lambda_1, \log_e \lambda_2)$ is an equivalence range for Δ, and *vice versa*. Because it does not matter which of the two vaccines is called 'vaccine 1' and which is called 'vaccine 0', the parameter of interest can be both θ or $1/\theta$. For this reason, an equivalence range for θ is usually defined as $1/\lambda$ to λ. This corresponds to setting the equivalence range for Δ to the symmetrical range $-\log \lambda$ to $\log \lambda$. An often used equivalence range for the ratio θ advised by FDA/CBER is 0.67 to 1.5 [30, 31].

Example 6.1. Joines and colleagues report the results of a combination vaccine trial [32]. They compared a combination hepatitis A and B vaccine with the monovalent vaccines. Both infectious diseases can be fatal. The major cause of hepatitis A is ingestion of faecally contaminated food or water. Hepatitis B is a sexually transmitted disease. In the trial, 829 adults were randomized to receive either the combination vaccine by intramuscular injection in the deltoid on a 0-, 1- and 6-month schedule, or separate intramuscular injections in the deltoids of opposite arms with hepatitis A vaccine on a 0 and 6 months schedule and hepatitis B vaccine on a 0, 1 and 6 months schedule. The primary analysis was a noninferiority analysis for the incidences of severe soreness, a safety endpoint. The secondary analysis was an equivalence analysis for the seroconversion rates for hepatitis A and the seroprotection rates for hepatitis B. In this example, the focus will be on the secondary analysis, the analysis of the month 7 immunogenicity data. Antibody titres to hepatitis A (anti-HAV) were determined using an enzyme immunoassay kit, and seroconversion for hepatitis A was defined as an anti-HAV titre ≥ 33 mIU/ml. Antibody titres to hepatitis B (anti-HBs) were determined using a radioimmunoassay kit, and seroprotection for hepatitis B was defined as an anti-HBS titre ≥ 10 mIU/ml. In total, 533 subjects were included in the per-protocol sample, 264 vaccinated with the combination vaccine and 269 vaccinated with monovalent vaccines. The main reason for exclusion from the per-protocol sample was being seropositive to hepatitis A or hepatitis B at baseline.

At month 7, the anti-HBs geometric mean titre and geometric standard deviation were 2,099 and 6.8 in the combination vaccine group and 1,871 and 9.5 in the monovalent vaccines group. Hence, the geometric mean ratio was

$$GMR = 2,099/1,871$$
$$= 1.12.$$

Because the geometric mean titre was not a secondary outcome, no equivalence range for the ratio θ was specified. For illustrative purposes, here, the range 0.67 to 1.5 will be used. The arithmetic means of the log-transformed antibody titres were

$$\log_e 2,099 = 7.65 \quad \text{and} \quad \log_e 1,871 = 7.53.$$

The corresponding standard deviations were

$$\log_e 6.8 = 1.92 \quad \text{and} \quad \log_e 9.5 = 2.25$$

giving as value for the pooled standard deviation 2.09. Standard error (6.2) takes the value

$$2.09\sqrt{1/264 + 1/269} = 0.18.$$

Thus, the value for the test statistic is

$$Z_{EQ} = \frac{0.41 - |7.65 - 7.53|}{0.18}$$
$$= 1.61,$$

where $0.41 = \log_e 1.5$. The corresponding P-value, from the t distribution with $(264 + 269 - 2) = 531$ degrees of freedom, is 0.054. Thus, the null hypothesis that the combination vaccine and the monovalent vaccine are not equivalent with respect to the induced anti-HBs responses cannot be rejected.

The 97.5th percentile of the central t distribution with 531 degrees of freedom is 1.964. Hence, the lower and upper bounds of the two-sided 95% confidence interval for Δ are

$$(7.65 - 7.53) - 1.964(0.18) = -0.23$$

and

$$(7.65 - 7.53) + 1.964(0.18) = 0.47.$$

By taking the antilogs of these limits, the limits of the 95% confidence interval for the geometric mean ratio θ are obtained. The resulting interval $(0.79, 1.60)$ does not fall in the equivalence range 0.67 to 1.5.

Suppose that another secondary objective was to demonstrate that the combination vaccine is noninferior to the monovalent vaccine with respect to the induced anti-HAV responses. The month 7 anti-HAV geometric mean titre and geometric standard deviation were 4,756 and 3.1 for the combination vaccine group and 2,948 and 2.5 in the monovalent vaccines group. Here, $GMR = 1.61$. The arithmetic means of the log-transformed antibody titres were

$$\log_e 4{,}756 = 8.47 \quad \text{and} \quad \log_e 2{,}948 = 7.99$$

with standard deviations

$$\log_e 3.1 = 1.13 \quad \text{and} \quad \log_e 2.5 = 0.92,$$

respectively. Here, the pooled standard deviation is 1.03, with $SE_P(D_{10}) = 0.09$. The lower bound of the one-sided 97.5% confidence interval for Δ is

$$(8.47 - 7.99) - 1.964(0.09) = 0.30,$$

which corresponds to a lower bound of

$$e^{0.30} = 1.35$$

for the geometric mean ratio θ. Because the lower bound falls to the right of the non-inferiority bound 0.67, the null hypothesis that the combination vaccine is inferior to the monovalent vaccine can be rejected.

6.4 Equivalence and Noninferiority Trails with a Seroresponse Rate as Outcome

Equivalence or noninferiority of two vaccines with a seroprotection or a seroconversion rate (i.e., a proportion) as endpoint is usually demonstrated by means of the confidence interval approach, with the risk difference as the parameter of interest. In case of nonsmall group sizes, the asymptotic confidence interval based on the Wilson method can be used (Sect. 3.5.2), while in case of small group sizes the exact interval based on the Suissa and Shuster exact test (Sect. 3.5.3) should be used. Mostly used equivalence and noninferiority margins for the risk difference are -0.05 (-5%) and -0.10 (-10%).

Example 6.1. (continued) The equivalence of the combination vaccine and the monovalent was to be demonstrated using the seroconversion rates for hepatitis A and the seroprotection rates for hepatitis B. For hepatitis A, equivalence was to be concluded if the two-sided 95% confidence interval for the seroconversion risk difference would fall in the equivalence range -0.05 to 0.05. The observed seroconversion rates were

$$267/269 = 0.993 \quad \text{and} \quad 263/264 = 0.996$$

for the combination vaccine group and the monovalent vaccines group, respectively. The estimated risk difference thus was

$$0.993 - 0.996 = -0.003.$$

The 95% confidence interval for the risk difference, $(-0.023, +0.014)$, is contained in the pre-defined equivalence range, and equivalence can be concluded.

With SAS, it is possible to test the two one-sided hypotheses associated with proving equivalence using the Wilson-type test statistic on the left-hand side of equation (3.7). The SAS code is

SAS Code 6.1 *TOST Procedure for Rates using the Wilson-Type Test Statistic*

```
proc freq;
  table vaccine*seroconverted /
    riskdiff (equivalence method=fm margin=0.05) alpha=0.025
  run;
```

The TOST P-values for the example above are 0.0004 and <0.0001. Both null hypotheses can be rejected. Note that the code can also be used to find the Wilson-type confidence limits by trial and error. The lower confidence limit is the smallest value for the lower margin for which the corresponding TOST P-value is ≥ 0.025. For `margin=0.022` the P-value is 0.0306, for `margin=0.023` the P-value is 0.0262, and for `margin=0.024` the P-value is 0.0224. From this, it can be concluded that the lower limit of the Wilson-type confidence limit is -0.023.

6.5 Vaccine Lot Consistency Trials

Both FDA/CBER and the European Medicines Agency (EMA) require, prior to licensure of a vaccine, proof that the vaccine production process is stable and that consistent lots can be produced. As part of this requirement a clinical study must be performed, a so-called lot consistency trial. The objective of a vaccine lot consistency trial is to show that the, preferably consecutively produced, lots (batches) are similar with respect to the induced immunogenicity. Subjects are randomly assigned to be vaccinated with vaccine from one of three lots. The post-vaccination blood samples of the subjects are assayed, and the antibody values are compared between the three lots. Lot consistency is concluded if all three pair-wise post-vaccination geometric mean ratios are close to one. Vaccine lot consistency trials are thus an example of equivalence studies.

6.5.1 Lot Consistency and the Confidence Interval Method

The most frequently applied method to demonstrate lot consistency is to calculate two-sided $100(1-\alpha)\%$ confidence intervals for the three pair-wise geometric mean ratios. If all three confidence intervals fall within the pre-defined equivalence range, then lot consistency is concluded.

Example 6.2. Nauta discusses the statistical analysis of influenza vaccine lot consistency trials [33]. The example he uses is a lot consistency trial with a virosomal influenza vaccine. (Virosomes are haemagglutinin and neuraminidase antigens linked to globular lipid membranes (liposomes), which are believed to have an adjuvant effect.) Following ICH guideline E9, the data of this equivalence study were analyzed according to the per-protocol principle [34]. Prior to unblinding of the database, the Blind Review Committee excluded 10 (2.7%) of the 373 randomized subjects from the per-protocol sample. Anti-HA antibody titres were determined using the haemagglutination inhibition (HI) test. As log transformation, the standard log transformation with $D = 10$ was used. The equivalence range for the pair-wise geometric mean ratios was predefined to be 0.35–2.83. In Table 6.1, the results for the A-H3N2 strain are summarized. For lot #2 versus lot #1, the pooled standard deviation is 1.585. The 97.5th percentile of the t distribution with

Table 6.1 Summary statistics of an influenza vaccine lot consistency trial (A-H3N2 strain, per-protocol sample)

	lot#1 ($n = 123$)	lot#2 ($n = 123$)	lot#3 ($n = 117$)
Geometric mean titre	192.9	162.2	202.5
Arithmetic mean*	5.27	5.02	5.34
Standard deviation*	1.57	1.60	1.57

*log-transformed antibody titres

$(123 + 123 - 2) = 244$ degrees of freedom is 1.970. Hence, the lower and the upper bound of the two-sided 95% confidence interval for Δ are

$$(5.02 - 5.27) - 1.970\sqrt{1.585(2/123)} = -0.648$$

and

$$(5.02 - 5.27) + 1.970\sqrt{1.585(2/123)} = 0.148.$$

By taking the antilogs (to the base 2), the two-sided 95% confidence intervals for the geometric mean ratio for lot #2 versus lot #1 is obtained:

$$(0.638, 1.108).$$

The two-sided 95% confidence intervals the geometric mean ratios for lot #3 versus lot #1 and for lot #3 versus lot #2 are

$$(0.796, 1.384) \quad \text{and} \quad (0.944, 1.651).$$

All three confidence intervals fall in the pre-defined equivalence range 0.38 to 2.83, and for the A-H3N2 strain lot consistency can be concluded.

6.5.2 The Wiens and Iglewicz Test to Inspect the Consistency of Three Vaccine Lots

Proving lot consistency can also be formulated as a hypothesis testing problem. Wiens and Iglewicz developed a statistical test to demonstrate the equivalence of three treatments [35]. The *Wiens and Iglewicz test*, which requires Normal data, can be used to demonstrate lot consistency. Let Δ_{ij} denote the difference between the expected means of the log-transformed antibody values of the ith and jth lot. To demonstrate lot consistency, the null hypothesis

$$H_0 : \max\{|\Delta_{ij}| \geq \delta\}$$

is tested against the alternative

$$H_1 : \max\{|\Delta_{ij}| < \delta\}$$

with δ the equivalence margin. Wiens and Iglewicz propose to test this null hypothesis by evaluating Z_{EQ} in (6.1) for all three pair-wise comparisons, and then to use the following min test statistic:

$$Z_{\min} = \min \left\{ \frac{\delta - |D_{ij}|}{SE(D_{ij})} \right\},$$

where D_{ij} is the difference between the arithmetic sampling means of the jth and the ith lot and $SE(D_{ij})$ the standard error of this difference. By definition, $Z_{\min}$ is a one-sided test, and the null hypothesis is rejected for large values for $Z_{\min}$. To test the above overall null hypothesis at the $\alpha/2$ significance level, the value of $Z_{\min}$ can be compared with the $100(1-\alpha/2)$th percentile of the standard Normal distribution.

Example 6.2. (continued) For the A-H3N2 strain, the observed differences are

$$D_{12} = -0.25, D_{13} = 0.07 \quad \text{and} \quad D_{23} = 0.32.$$

The standard errors of these differences are

$$SE(D_{12}) = \sqrt{1.57^2/123 + 1.60^2/123} = 0.202$$
$$SE(D_{13}) = \sqrt{1.57^2/123 + 1.57^2/117} = 0.203$$
$$SE(D_{23}) = \sqrt{1.60^2/123 + 1.57^2/117} = 0.205.$$

This gives (with $\delta = \log_2 2.83 = 1.5$)

$$Z_{\min} = \min \left\{ \frac{1.5 - 0.25}{0.202}, \frac{1.5 - 0.07}{0.203}, \frac{1.5 - 0.32}{0.205} \right\}.$$

The resulting value, 5.76, is highly statistically significant.

General properties of the min test were already discussed in Sect. 3.8. A nice property of the test is that it is an overall test, and a correction for multiplicity is not needed. The drawback of the test is the complexity of its sampling distribution, which makes it difficult to evaluate the type I error rate. Wiens and Iglewicz have studied the type I error rate for $Z_{\min}$ [35]. Because lots can be re-ordered, without loss of generality

$$0 \le \Delta_{12} \le \Delta_{13}.$$

Let

$$\Delta = \Delta_{13} \quad \text{and} \quad \rho = \Delta_{13}/\Delta_{12},$$

with σ_i and n_i the within-lot standard deviation of the log-transformed antibody values and the sample sizes of the ith lot, and let

$$SE_{ij} = \sqrt{\sigma_i^2/n_i + \sigma_j^2/n_j}.$$

The type I error rate of Z_{min} depends on Δ/SE_{ij} and ρ. To shed light on this, in Table 6.2 Monte Carlo simulation results are presented, for the simple case that

$$\sigma_1 = \sigma_2 = \sigma_3 \quad \text{and} \quad n_1 = n_2 = n_3 = n.$$

The type I error rate is defined as

$$\Pr(Z_{min} > z_{1-\alpha/2}|\Delta = \sigma).$$

(The algorithm used for the simulation was similar to the algorithm for sample size estimation for lot consistency trials outlined in Sect. 6.7.3). The actual type I error rates are lowest for ρ close to 0.0 or 1.0 and highest for $\rho = 0.5$, and then they are close to the nominal error rate for nonsmall n and δ/σ.

Wiens and Iglewicz show that when

$$(\delta/\sigma)\sqrt{n/2} > 5.0$$

the actual type I error approaches the nominal one for $\rho = 1.0$. This is the same as requiring that

$$n > \frac{50}{(\delta/\sigma)^2}. \tag{6.4}$$

Equation (6.4) can be thus used to decide if the lot sample sizes guarantee that the actual type I error rate of the trial is sufficiently close to the nominal level. If $\delta/\sigma = 1.0$, then n must be greater than 50, if $\delta/\sigma = 0.5$, then n must be greater than 200 and when $\delta/\sigma = 0.25$, then n must be greater than 800. The simulation results in Table 6.2 are consistent with this.

Example 6.2. (continued) If it is assumed that $\sigma_1 = \sigma_2 = \sigma_3 = 1.6$, then $\delta/\sigma = 1.5/1.6 = 0.94$. To secure a nonconservative actual type I error rate, the group size per lot should be at least

Table 6.2 Actual type I error rates in lot consistency trials, for $\alpha = 0.05$ and with as critical value $z_{1-\alpha/2}$

δ/σ	n	0	0.25	ρ 0.5	0.75	1
0.25	500	0.011	0.029	0.044	0.033	0.011
	300	0.013	0.031	0.033	0.024	0.013
	100	0.001	0.001	0.002	0.002	0.002
0.5	500	0.011	0.050	0.048	0.047	0.010
	300	0.013	0.041	0.045	0.042	0.012
	100	0.010	0.032	0.042	0.033	0.015
1.0	500	0.013	0.048	0.049	0.053	0.014
	300	0.012	0.045	0.050	0.051	0.009
	100	0.015	0.044	0.048	0.044	0.015

$$\frac{50}{0.94^2} = 57,$$

which was the case.

The Wiens and Iglewicz approach and the confidence interval approach yield near similar results. The two approaches differ only in the standard errors being used and the critical value (derived from the standard Normal distribution versus the t distribution).

6.6 Discussion

Equivalence and noninferiority trials are not uncontroversial. There are several reasons for this. One, perhaps the most important, reason is that it is often difficult to justify the choice of the margin. An overlay strict margin will require a prohibitively large sample size, while a too large margin will not be clinically meaningful. (That a margin leads to a too large sample size is, admittedly, not a very strong argument. Indeed, what matters is the clinical relevance of the differences that the margin allows.) Consider the problem of deciding an equivalence range for the geometric mean ratio. The range 0.67 to 1.5 is generally considered to be a reasonable one, not too wide but also not too strict. A proper justification for this range would take into account the strength of the relationship between the antibody measurements and the probability of clinical protection from infection. If the protection curve (see Sect. 8.2) would be a steep one, then a very small range would be appropriate, while a less steep one would allow a broader range. The difficulty is that in practice the relationship is seldom known with sufficient detail.

Many vaccines contain antigen of more than one serotype. In that case, equivalence or noninferiority must usually be demonstrated for all serotypes. A much applied approach is to demonstrate equivalence or noninferiority at the $\alpha/2$ significance level for each of the serotypes. The intersection-union (IU) principle then allows that equivalence or noninferiority is claimed on vaccine level, and no multiplicity correction is needed. This approach is known to be conservative under many circumstances. Kong, Kohberger and Koch have proposed *min* tests for vaccine equivalence and noninferiority trials with multiple serotypes [36, 37]. These tests take the correlation between the endpoints into account. Simulation results, however, show that for trials with multiple binomial endpoints the *min* test leads to only modest increases in power.

A statistical novelty of recent date is simultaneous testing of noninferiority and superiority [38]. Basically, if in a noninferiority trial the null hypothesis is rejected, one can proceed to test the null hypothesis for superiority. In a superiority trial, if the null hypothesis is not rejected, one can proceed with a test for noninferiority. The strategy can be justified by either the IU principle or the closed testing principle, and no multiplicity adjustment is needed [39, 40]. The strategy is not undisputed. One of its critics, Ng, argues that the strategy allows an investigational treatment to claim superiority by chance alone without risking the noninferiority claims, which will increase the number of erroneous claims of superiority [41]. Despite this criticism, the strategy has quickly found its way into clinical development. Leroux-Roels and co-workers compared an intradermal (injected between the layers of the skin) trivalent inactivated split-viron influenza vaccine with an intramuscular control vaccine [42]. They conclude that the intradermal vaccine induces noninferior humoral immune responses against all three virus strains included in the vaccines, because all three two-sided 95% confidence intervals for the geometric mean ratios fell above the pre-specified noninferiority margin. In addition, they conclude superior responses against both A strains, because for these two strains the confidence interval fell above 1.0. This second conclusion is disputable. In the Statistical Methods section, the authors explain that noninferiority was tested at vaccine level (i.e., to be demonstrated for *all* strains contained in the vaccines), but that superiority was tested per strain, at the two-sided significance level 0.05. As argued above, because noninferiority was to be shown at vaccine level a multiplicity correction was not needed. But because superiority could have been claimed for any number of strains, here a multiplicity correction should have been applied. Thus, contrary to what was claimed, the statistical analysis does not warrant the conclusion of a superior response against both A strains.

6.7 Sample Size Estimation

6.7.1 Comparing Two Geometric Mean Responses

The statistical power of a clinical vaccine trial with the primary objective to demonstrate that the immunogenicity induced by an investigational vaccine is equivalent or noninferior to that induced by a control vaccine, and with either a geometric mean titre or a geometric mean concentration as outcome is, apart from the sample size and the significance level, dependent upon:

1. The equivalence/noninferiority margin $-\delta$.
2. The within-group standard deviation σ of the log-transformed immunogenicity values.
3. The difference $\Delta = \mu_1 - \mu_0$, with μ_1 and μ_0 the arithmetic means of the probability distributions underlying the log-transformed immunogenicity values of the investigational and the control vaccine group.

As will be shown below, the statistical power is profoundly sensitive to assumptions about Δ, which makes sample size estimation for vaccine equivalence and noninferiority trials a challenging exercise. To power trials on the assumption of a zero difference between the vaccines is therefore not recommended. Better is to assume some amount variation between the vaccines.

Under the usual assumption that the log-transformed immunogenicity values are Normally distributed, the statistical power of equivalence and noninferiority trails should be estimated from noncentral t distributions. For equivalence trials sample size Formula (5.5) in the book by Julious applies [20]. The formula can be solved with the procedure POWER.

Example 6.1. (continued) An investigator wishes to know the statistical power of the trial for the combination vaccine versus the monovalent hepatitis B vaccine. The equivalence margin for Δ is $\delta = \log_e 1.5 = 0.41$. Suppose that for log-transformed anti-HBs titres the convention is to set σ to 2.0. For a first sample estimate Δ is set to 0.0. The desired statistical power is 0.90. The required sample size can be calculated with the following SAS code.

SAS Code 6.2A *Sample Size Estimation for an Equivalence Trial with Log-Transformed Immunogenicity Data Assuming a Zero Mean Difference*

```
proc power;
   twosamplemeans test=equiv_diff alpha=0.025
   meandiff=0   stddev=2.0
   lower=-0.41  upper=0.41
   power=0.9    npergroup=.;
run;
```

When this code is run, a required sample size of 620 subjects per group is found. Next, the investigator wishes to study the robustness of this estimate when it is assumed that the combination vaccine is somewhat less immunogenic than the monovalent vaccine, i.e., when it is assumed that difference Δ is less than zero or, which is the same, the geometric mean ratio θ is less than one. He proposes two values for the ratio, $\theta = 0.95$ and $\theta = 0.90$. These values correspond to $\Delta = -0.051$ and $\Delta = -0.105$. The robustness of the above sample size estimation can be inspected with the following SAS code:

SAS Code 6.2B *Statistical Power of an Equivalence Trial with Log-Transformed Immunogenicity Data Assuming a NonZero Mean Difference*

```
proc power;
   twosamplemeans test=equiv_diff alpha=0.025
   meandiff=0, -0.051, -0.105
   stddev=2.0
   lower=-0.41   upper= 0.41
   npergroup=620 power=.;
run;
```

The SAS output is given below.

SAS Output 6.2B

```
                The POWER Procedure
         Equivalence Test for Mean Difference
               Fixed Scenario Elements

     Distribution                        Normal
     Method                               Exact
     Lower Equivalence Bound              -0.41
     Upper Equivalence Bound               0.41
     Alpha                                0.025
     Standard Deviation                       2
     Sample Size Per Group                  620

                Computed Power

                  Mean
        Index     Diff      Power
          1       0.000     0.900
          2      -0.051     0.866
          3      -0.105     0.760
```

If θ is assumed to be 0.9 instead of 1.0, the statistical power drops from 0.90 to 0.76. This demonstrates the critical dependency of the statistical power on Δ. If the value of 0.9 for θ would be considered to be more likely, then a sample size of 906 subjects per group would be required to be secured of a statistical power of 0.9.

The formula for the statistical power of a noninferiority trial with Normal data is Formula (6.6) in the book by Julious [20]. This formula can be evaluated with either SAS code 6.2A or SAS code 6.2B, with the variable upper set to a very large positive value, say 999.

Example 6.1. (continued) Suppose that the investigator also wishes to know the statistical power of the trial for combination vaccine versus the monovalent hepatitis A vaccine. The within-group standard deviation σ is set to 1.25, Δ to 0.0 and the noninferiority margin to -0.41. First it is assumed that Δ equals zero, and the investigator would like to know the required numbers of subjects to be secured of a statistical power of 0.9.

SAS Code 6.3 *Statistical Power of a NonInferiority Trial with Log-Transformed Immunogenicity Data, Assuming a Zero Mean Difference*

```
proc power;
    twosamplemeans test=equiv_diff alpha=0.025
    meandiff=0   stddev=1.25
    lower=-0.41 upper=999
    power=0.9    npergroup=.;
run;
```

If this code is run, a sample size of 197 subjects per group is found.

6.7.2 Comparing Two Seroresponse Rates

Sample size estimation for a noninferiority trial comparing two proportions and with the risk difference as effect measure is discussed in Sect. 11.3.1 of the book by Julious [20]. Sample sizes from the preferred method, Method 1 (based on expected rates, the least conservative of the three methods), can be computed with procedure POWER. With this procedure, sample sizes can be calculated for testing null hypotheses of the form

$$H_0 : \tilde{\pi}_B - \tilde{\pi}_A \leq -d$$

In case of a noninferiority trial, the null hypothesis to test is

$$H_0 : \pi_1 - \pi_0 \leq -\delta.$$

Sample sizes for this null hypothesis can thus be estimated by setting $\tilde{\pi}_A$ to π_0, $\tilde{\pi}_B$ to π_1 and $-d$ to $-\delta$.

An investigator wishes to estimate the sample size corresponding to a power of 0.9 for a noninferiority trial with expected seroprotection rates $\pi_1 = 0.70$ and $\pi_0 = 0.75$ and noninferiority margin $\delta = -0.1$, and with the null hypothesis being tested at the one-sided significance level 0.025. The required sample size can be obtained with the following SAS code:

SAS Code 6.4 *Sample Size Calculation for a NonInferiority Trial with a Seroprotection or a Seroconversion Rate as Outcome*

```
proc power;
    twosamplefreq test=pchi
    alpha=0.025 sides=1
    groupproportions=(0.75 0.70)
    nullpdifference=-0.10
    power=0.9 npergroup=.;
run;
```

The computed sample size is 1,674 subjects per group.

A formula for sample size estimation for an equivalence trial comparing two proportions and with the risk difference as effect measure is Formula (12.4) in the book by Julious [20]. SAS procedure POWER does not contain a feature for equivalence trials with a rate as endpoint. But the formula is easy to program.

6.7.3 Lot Consistency Trials

The standard approach to estimate the statistical power of a lot consistency design is to estimate separately the power of each of the three pair-wise comparisons, and then combine the resulting estimates to obtain an overall estimate.

Example 6.4. Ganju, Izu and Anemona investigate sample size estimation for vaccine lot consistency trials [43–45]. The example they use is a lot consistency trial for a quadrivalent vaccine for the prevention of meningococcal disease caused by *N. meningitidis* serogroups A, C, Y and W-135. For the C serogroup, they assume that $\sigma = \sqrt{(6.15)}$ for the $\log_2$-transformed antibody titres. They find that if there is no between-lot variation, i.e., if it is assumed that $\Delta_{12} = \Delta_{13} = 0$, that to be secured of an overall statistical power of 0.9, a sample size of $n = 500$ subjects per lot would be required (for $\alpha = 0.05$, and 1/1.5 to 1.5 as equivalence range). Because of the assumption that $\Delta_{12} = \Delta_{13} = 0$, each of the three single pair-wise comparison have the same power. This power can be calculated using SAS code 6.2A with

```
meandiff=0      stddev=2.480
upper=0.585     lower=-0.585
alpha=0.05      npergroup=500
```

with $0.585 = \log_2 1.5$. The computed power for the single pair-wise comparisons is 0.963. Using the inequality (3.15), it follows that the overall statistical power for this lot consistency design is ≥ 0.889. If independence of the three pair-wise comparisons is assumed, then the overall statistical power is $0.963^3 = 0.893$.

Ganju and colleagues show that assumptions about Δ_{12} and Δ_{13} can have a profound impact on the statistical power of the design. They strongly argue against assuming that the between-lot variation is zero, and they advise to assume some amount of variation between lots. If nonzero between-lot variation is assumed, i.e., if it is assumed that $\Delta_{13} > 0$, then the statistical power will be highest for $\Delta_{12} = \Delta_{13}/2$ and lowest for $\Delta_{12} = 0$. The explanation for this is that equally spaced means are less variable than unequally spaced means and hence have a greater probability of demonstrating consistency.

Example 6.4. (continued) If Δ_{13} is set to 0.1 and Δ_{12} to 0.06, then with a lot size of 500 the power of the three pair-wise comparisons are: 0.949 (lot #1 versus lot #2), 0.923 (lot #1 versus lot #3) and 0.956 (lot #2 versus lot #3). These estimates can be obtained using in SAS code 6.2B the statement

```
meandiff=0.1, 0.06, 0.04
```

Thus, now the overall statistical power would be

$$\geq (0.923 + 0.949 + 0.956) - 2 = 0.828,$$

or 0.837 if independence is assumed.

The standard approach to estimate the overall statistical power of a lot consistency design is slightly flawed, because it does not take the correlation between the three pair-wise comparisons into account. That the correlation between the comparisons is not zero is easy to see. If the mean differences between (a) lot #1 and lot #2 and (b) lot #1 and lot #3 are known, then the mean difference between lot #2 and lot #3 is also known. Thus, the assumption that the three pair-wise comparisons are independent does not hold. Second, because the comparisons are dependent, separate estimation of the power of each of the three pair-wise comparisons also introduces bias. In practice, however, the amount of bias will be negligible, and the standard approach will give a good approximation of the actual power of the design.

An alternative method to estimate the overall power is Monte Carlo simulation. The algorithm for the simulation is as follows. A large number ($\geq 5,000$) of trials is simulated. Per trial three random samples of size n are generated, with the data of the ith sample representing the log-transformed antibody values of the ith lot. The first random sample is drawn from a $N(0, \sigma^2)$ distribution, the second from a $N(\Delta_{12}, \sigma^2)$ distribution and the third from a $N(\Delta_{13}, \sigma^2)$ distribution. For each trial, the confidence intervals for the three pair-wise geometric mean ratios are calculated, and the trial is declared significant if all three confidence intervals fall in the pre-defined equivalence range. The statistical power of the design is then estimated as the proportion of simulated trials yielding a significant result. With this approach, the correlation between the comparisons is taken into account. For the example above, the simulated overall statistical power is 0.887.

Chapter 7
Vaccine Field Efficacy Trials

7.1 Introduction

Vaccines can produce different kinds of effects, which can be at subject level or at population level. Halloran, Struchiner and Longini present a theoretical framework for vaccine effects and designs of vaccine field efficacy trials for the estimation of these effects [46, 47]. The nomenclature here has been taken from Chap. 2 of their book on vaccine efficacy studies [48].

At subject level, historically the effect of interest has been the *vaccine efficacy for susceptibility* to infection. Here, the question is how well the vaccine protects vaccinated subjects against infection or disease, i.e., to what degree vaccination reduces the probability that a subject becomes infected or diseased if exposed to the pathogen. Standard endpoints in vaccine field efficacy trials for susceptibility to infection are occurrence of infection or disease and time to infection or disease.

Often, infection confers lifelong protection against the disease. This is for instance the case for mumps, measles and hepatitis A. For these diseases, a subject can get infected only once. This is, however, not true for all infectious diseases. Examples of diseases with possibly recurrent infections are acute otitis media (middle ear infection), genital herpes, meningitis and cystitis (an inflammatory disorder of the bladder). The reason why infection does not lead to lifelong protection is usually either that the naturally acquired antibodies against the pathogen do not offer sufficient protection, or exposure to serotypes of the pathogen that are not recognized by the antibodies. Cystic fibrosis (CF) is an inherited disease of the mucus. An abnormal gene causes the mucus to become thick and sticky. The mucus builds up in the lungs and blocks the airways, which makes it easy for bacteria to grow, leading to repeated, life-threatening lung infections. Over time, these infections can cause chronic progressive pulmonary disease, the most frequent cause of death in CF patients. The most prevalent of these infections is the one caused by the bacterium *Pseudomonas aeruginosa*. Naturally acquired antibodies against the bacterium often do not offer sufficient protection against the very virulent infections. A CF patient can become chronically colonized, with subsequent resistance to antibiotic courses. In a trial with a *P. aeruginosa* vaccine in CF patients measures of effect may thus be: time to initial infection, time to colonization with *P.*

aeruginosa, becoming chronically infected, number of recurrent infections or time between subsequent infections.

Vaccine efficacy for progression or pathogenesis is the protection the vaccination offers once a person has become infected. The vaccine may increase the incubation period, i.e., the time between infection and disease. Other effects of interest could be to what degree the vaccine reduces the intensity, the duration or the mortality from disease. Halloran gives the example of human immunodeficiency virus (HIV). A HIV vaccine may reduce the post-infection *viral load*, the amount of virus in body fluids. In HIV, keeping the viral load level as low as possible for as long as possible decreases the complications of the disease and prolongs life.

Whereas vaccine efficacy for susceptibility to infection requires that the trial participants are free of infection at the time of their enrollment into the trial, vaccine efficacy for progression or pathogenesis can only be studied in infected subjects.

A vaccinated subject may be less infectious to others, or he or she may be infectious for a shorter period of time. This is called *vaccine efficacy for infectiousness*. These effects are of relevance at population level, because reduction of infectiousness has usually important health consequences. The effects will slow down the spread on the infection in the population. Vaccination of a large fraction of the population may lead to *herd immunity*. If a high percentage of a population immune is to an infection, then the spread of the infection may be prevented because it cannot find new hosts.

7.2 Some Critical Aspects of Vaccine Field Trials

7.2.1 Efficacy versus Effectiveness

In the literature, a distinction is made between vaccine field efficacy and vaccine field effectiveness trials. With *vaccine efficacy* is meant the degree in which the vaccine offers protection against the target infection or disease, influenza for example. With *vaccine effectiveness* is meant the degree in which the vaccine offers protection against diseases for which the subjects' susceptibility may be negatively influenced by the infection (complications following infection). This may specially be the case in subjects with a serious chronic illness. Chronic illnesses such as cardiovascular or pulmonary diseases, metabolic diseases (e.g., diabetes and renal dysfunction), or immunodeficiency (e.g., during or after treatment for cancer) increase the risk of complications following influenza infection. Conditions that impair the handling of respiratory secretions, such as CF, predispose to respiratory infections and also increase the risk of developing complications of influenza. So, in a vaccine field effectiveness trial, the primary endpoint measure can be exacerbation of an underlying disease (e.g., chronic airway obstruction), or hospitalization, or, indeed, death.

The efficacy or effectiveness of a vaccine will depend on the primary endpoint, i.e., on the case definition. It may be high for one endpoint but only moderate for another. Less understood is that it also depends on the sensitivity and specificity of the diagnostic test used and the definition of the surveillance period. The first to point this out were Orenstein and colleagues, who, on behalf of the United States Centers for Disease Control (CDC), published a lengthy paper on vaccine field efficacy trials [49]. They argued that in these trials specificity of the diagnostic test is usually more critical than sensitivity for assessing vaccine efficacy, and that surveillance period should be restricted to the peak outbreak period when disease incidence is highest.

7.2.2 The Influence of the Sensitivity and Specificity of the Diagnostic Test on the Vaccine Efficacy Estimate

One of the most critical aspects of vaccine efficacy trials is *case definition*. First, it has to be decided whether infection or disease is the endpoint of interest. This choice is of importance because infection does not necessarily imply developing the disease. As a rule of the thumb, vaccine efficacy trials with disease as endpoint require larger sample sizes and a longer surveillance period than trials with infection as endpoint. On the other hand, case finding may be easier with disease as endpoint. Disease is usually accompanied by specific clinical symptoms while infection requires laboratory confirmation. With infections with a long incubation period, such as HIV for example, infection as endpoint would require repeated laboratory testing, which sometimes is difficult to organize (trial participants having to visit the investigational site at pre-defined times, etc.) and can be very costly (while most laboratory results will be negative.) If the incubation period is short, like in the case of pertussis, laboratory testing is often done after observing clinical symptoms of the disease. A drawback of this case finding strategy is that *asymptomatic infections*, which are infections without clinical symptoms, go undetected. In case of influenza, culture confirmation is only possible during the first two to three days after infection. If the culture specimen collection is done too late, the infection may go undetected. It is sometimes argued that it is disease that matters, not infection. That is a too hasty conclusion. Infections do not only cause the disease, some can also do damage to organs. Asymptomatic infections should not be considered as being without risk. Sexually transmitted infections, in particular, are known for not producing clinical symptoms. If treatment for the infection is delayed or never given, this can cause permanent damage to the reproductive organs. In fact, almost any type of infection can impair fertility, in particular, those that affect the reproductive tract, including the prostate, the epididymis or the testis. A harmless infection such as the common cold may temporarily lower the sperm count.

Case finding requires a diagnostic test. (Diagnostic test is used here in the broadest sense. It can be a single clinical or laboratory test but it can also be a diagnostic strategy, e.g., laboratory testing only after the manifestation of certain clinical symptoms.) If the test misclassifies noncases as cases of the infection or the disease, then the vaccine efficacy estimate will be biased towards null. If, on the other hand, the test misclassifies cases as noncases, this will not necessarily bias the efficacy estimate, but it may. If the diagnostic test detects moderate to severe cases easier than mild cases, and if vaccinated cases are milder than placebo cases, this will result in a too high efficacy estimate.

Diagnostic tests are rarely totally accurate, and a proportion of the cases will be misclassified as noncases. Such cases are called *false-negative cases*. The proportion of misclassified cases will depend on a property of the diagnostic test known as the sensitivity. The *sensitivity* of a diagnostic test is the conditional probability that the test will be positive (Test+) if the disease is present (Disease+):

$$\text{sensitivity} = \Pr(\text{Test+} \mid \text{Disease+}).$$

By definition, the false-negative rate is equal to (1-sensitivity).

Not only may cases be misclassified as noncases, noncases may be classified as cases. Such cases are called *false-positive cases*, and their number will depend on the specificity. The *specificity* of a diagnostic test is the conditional probability that the test will be negative (Test−) given that the disease is absent (Disease−):

$$\text{specificity} = \Pr(\text{Test−} \mid \text{Disease−}).$$

The false-positive rate is equal to (1-specificity). As already noted, a low specificity of a diagnostic test will bias the vaccine efficacy estimate towards null. If the sensitivity of the clinical test is not perfect, this will not necessarily bias the estimate.

Example 7.1. Consider a pneumonia vaccine field efficacy trial. Diagnosis of pneumonia is difficult and there are numerous diagnostic tests: a clinical test with the use of the stethoscope, a chest x-ray or other imaging techniques, laboratory tests, such as sputum tests and blood tests, etc. Suppose that in a randomized trial, 500 elderly are vaccinated with a pneumonia vaccine and a further 500 with placebo, and that during the surveillance period (first year after vaccination) 25 subjects contract pneumonia, 5 in the vaccine group and 20 in the control group. In case of a placebo-controlled trial, vaccine efficacy for susceptibility is defined as the proportion of prevented cases (see Sect. 7.4.1). Thus, the vaccine efficacy for the prevention of pneumonia is

$$(20 - 5)/20 = 0.75.$$

Suppose that the sensitivity of the diagnostic test is not perfect, not be 1.0, but, say, 0.8. In that case, the (expected) number of false-negative cases would be

$$(1.0 - 0.8) \times 25 = 5,$$

with 4 misclassified cases in the placebo group and 1 in the placebo group. But the vaccine efficacy would still be correctly estimated:

$$(16 - 4)/16 = 0.75.$$

Next, suppose that not only the sensitivity but also the specificity would be less than 1.0, say, 0.95. The number of false-positive cases would be

$$(1.0 - 0.95) \times (1000 - 25) = 48.75,$$

i.e.,
$$(1.0 - 0.95) \times (500 - 20) = 24$$

misclassified cases in the placebo group, and

$$(1.0 - 0.95) \times (500 - 5) = 24.75$$

misclassified noncases in the vaccine group. The total number of cases in the placebo group would be
$$(16 + 24) = 40,$$

while the total number of cases in the vaccine group would be

$$(4 + 24.75) = 28.75.$$

And the vaccine efficacy would be estimated to be:

$$(40 - 24.75)/40 = 0.381,$$

which is indeed a bias towards null.

Finally, it may be that the diagnostic test is such that less severe cases go undetected, and that the less severe cases are predominantly in the investigational vaccine group. In that case, the vaccine efficacy will be overestimated.

Example 7.1. (continued) Suppose that the sensitivity of the diagnostic test for pneumonia is 1.0 for severe cases but only 0.5 for nonsevere cases, and that in the placebo group 30% of the cases are nonsevere but in the vaccine group 90%. In that case, in the placebo group

$$(1.0 - 0.5) \times 0.3 \times 20 = 3$$

cases will be misclassified, while in the vaccine group

$$(1.0 - 0.5) \times 0.9 \times 5 = 2.25$$

cases will be misclassified. The vaccine efficacy would be estimated to be

$$\frac{(20 - 3) - (5 - 2.25)}{20 - 3} = 0.838,$$

a modest overestimation of the vaccine efficacy.

7.2.3 Surveillance Period

In case of a diagnostic test based on clinical symptoms, it can be that the specificity of the test is dependent on the incidence of the infectious disease in the community. If the incidence increases, then the probability that the clinical symptoms are associated with the disease will also increase, in which case the false-positive rate will decrease and the specificity increase. This is why it is often advised to restrict the *surveillance period*, the period during which trial participants are followed up to see if they get infected or develop the disease, to the period of peak activity, when the infectious disease incidence is highest.

A surveillance period is called fixed if both the start and the length of the surveillance period are fixed. The surveillance period can be the same period for trial participants, the infectious disease season for example, usually a specified calendar period, e.g., October – March. But the surveillance period not need to be the same calendar period for all subjects. The surveillance period could, for example, be the first year after vaccination. But the surveillance period can also be a specified age period, e.g., between 3 and 12 months after birth. All these surveillance periods have in common that the start and the length is fixed.

7.3 Incidence Measures for Infection

The concept to quantify the occurrence of cases of a disease or infection is incidence. Incidence concerns new cases of the disease in a group of subjects who are initially disease-free. (The proportion of existing cases of the disease in the population is called the prevalence.) There are three incidence measure: the cumulative incidence, the incidence rate and the hazard rate. In a vaccine field efficacy trial the endpoint can be either infection or disease. For convenience, in this and the next sections it will be assumed that the endpoint is infection. If in a vaccine field efficacy trial the endpoint is infection then the three incidence measures are usually termed the attack rate, the infection rate and the force of infection.

7.3.1 Attack Rate

The *attack rate* is the risk an infection-free subject gets infected during a fixed surveillance period. The attack rate is estimated by the number of infected cases occurring during the surveillance period, divided by the total number of initially infection-free subjects. Attack rates depend upon the length of the surveillance period, the longer the period, the higher the attack rate.

To be able to estimate the attack rate, all trial participants must be followed up for the entire surveillance period (or until the time of infection). In practice, this is often difficult to achieve, due to the phenomenon of drop-out. For the drop-outs, the duration of the follow-up will be less than the planned length of the surveillance period, and the endpoint – infected yes/no – will be missing. There is no straightforward method to handle these missing data. A very crude approach is to perform a complete-case analysis, complemented by a worst-case sensitivity analysis in which all subjects with a missing endpoint are counted as infected if in the investigational group and as noninfected if in the control group. Vaccine field efficacy trials comparing infection rates or force of infection functions are much more flexible with respect to the handling of drop-outs.

7.3.2 Infection Rate

The *infection rate* is the risk of experiencing an infection during a given time unit, e.g., a month or a year. The infection rate depends on the chosen time unit. An infection rate of 0.001 per month corresponds to an infection rate of 0.012 per year. A condition underlying the concept of infection rate is that the risk of infection must be constant over time. It must be stressed that the condition of a constant risk over time is a very strong one. The general consensus is that the condition holds for many infectious diseases and many vaccines. But the infectious disease should not be a seasonal one, in which case the condition is not met. An example of a seasonal infectious disease is influenza. The risk of infection will be close to zero at the start and the end of the season, and it will be highest during the peak of the season. Seasonal change in the risk of infection is an often seen phenomenon in both temperate and tropical climates. If the protection a vaccine affords wanes over time then, depending on the length of the surveillance period, the condition of a constant risk may also not hold.

Because the risk is assumed to be constant over time, varying start times nor lengths of surveillance period per subject matter. This is an important difference with vaccine field efficacy trials comparing two attack rates, and it allows considerable flexibility in trial participant enrollment.

The infection rate is estimated by the *events-per-person-time statistic*, i.e., the number of infected cases s divided by the total person-time T

$$EPPT = s/T.$$

Consider a group of n initially infection-free subjects. One subject could enter the study 1 week after being vaccinated and be followed up for 6 months, while another subject could enter 5 weeks after being vaccinated and be followed up for 3 months, etc. Let T be the total person-time for the group, i.e., the duration of the infection-free period (or surveillance period if infection did not occur) for the first subject plus the duration of the infection-free period for the second subject, etc. The infection rate is then estimated by the number of infected cases divided by T. A total person-time of 5 years can be the resultant of 5 subjects each with an individual person-time of 1 year, or of 60 subjects each with an individual person-time of 1 month, or of n subjects all with a different individual person-time, some short, some long. For drop-outs, the duration of the follow-up can be set to the length of the time interval between the start of the surveillance period and the moment of drop-out. This requires the assumption of noninformative drop-out, meaning that the reason for drop-out is not related to the endpoint, occurrence of infection.

The usual assumption is that the number of cases is a realization from a Poisson distribution, meaning that there must be a uniform scatter of infected cases over time. A uniform scatter of cases over time is only possible if the infection rate for the (maximum) surveillance period is low. The reason for this is that the events-per-person-time estimator does not allow for the fact that in a clinical vaccine trial, during the surveillance period the number of subjects at risk for infection decreases. And if the number of subjects at risk decreases, so will the number of new cases. In that case the scatter over time will not be uniform and the Poisson assumption will not hold.

Finally, let $\pi(t)$ denote the attack rate for the surveillance period $[0, t]$ after vaccination, and $\lambda(t)$ the infection rate during a post-vaccination surveillance period of length t. If the risk of infection is constant over time and $\lambda(t)$ is low, then

$$\lambda(t) \approx \pi(t).$$

Thus, if the infection rate is low, then it will be approximately equal to the attack rate. This is of importance for the interpretation of vaccine efficacy estimators based on the events-per-person-time statistic.

7.3.3 Force of Infection

The third incidence measure is force of infection. Let $S(t)$ be the survivor function, i.e., the function giving the probability that a subject remains infection-free longer than some time t. The attack rate $\pi(t)$ of the infection for the surveillance period $[0, t]$ is

$$\pi(t) = 1 - S(t).$$

The *force of infection* $h(t)$ at t is defined as

$$h(t) = \frac{-\mathrm{d}S(t)/\mathrm{d}t}{S(t)},$$

the instantaneous change in the size of the infection-free population at time t divided by the size of the infection-free population at time t. To make this less abstract, imagine time to be discrete, then $h(t)$ would be the number of infected cases occurring at time t divided by the number of subjects being infection-free at $(t - 1)$. The force of infection $h(t)$ is thus indeed an incidence measure. It may be shown that

$$\pi(t) = 1 - \exp \int_0^t h(s)\mathrm{d}s.$$

A key difference with the infection rate is that the force of infection need not to be constant over time. In fact, no assumptions about $h(t)$ need to be made.

The length of the surveillance period need not to be fixed, it may vary from subject to subject. Data from drop-outs can be treated as censored observations, but here also, the reason for drop-out should not be related to the occurrence of infection. A second key difference with the infection rate is that here t measures the time since the start of the surveillance period, and this start must be fixed, e.g., the day of (the first) vaccination.

The force of infection function can be estimated using survival analysis techniques. If in procedure LIFEREG the option

```
plots = (h(name=force of infection));
```

is used, the force of infection function is plotted versus time.

7.4 Statistical Analysis of Vaccine Efficacy Data

Vaccine efficacy (VE) for susceptibility is generally defined as one minus a relative risk of infection parameter θ in the investigational vaccine group versus the control group:

$$VE = 1 - \theta. \qquad (7.1)$$

The relative risk of infection can be estimated by either the ratio of two attacks rates, the ratio of two infection rates, or the ratio of two force of infection functions. It will be shown that under the mild condition that the attack rate in the control group is small, independent of the relative risk estimator being used, the same vaccine efficacy parameter is estimated, namely

$$1 - \frac{\pi_1(t)}{\pi_0(t)}$$

with $\pi_1(t)$ and $\pi_0(t)$ the attack rate during the surveillance period $[0,t]$ in the investigational and the control group, respectively. In other words, the interpretation of vaccine efficacy is independent of the incidence measure being used.

Often it will not be sufficient to demonstrate that the vaccine efficacy is greater than zero, but that the requirement is that it has to be shown that the efficacy is substantially greater than zero. This is called *super efficacy*. For example, for influenza vaccines the FDA/CBER requirement is that the vaccine efficacy must be greater than 0.4 [30]. This is to be demonstrated by showing that the lower bound of the two-sided 95% confidence interval for the vaccine efficacy exceeds 0.4.

7.4.1 Comparing Two Attack Rates

If the surveillance period is fixed, then the relative risk in (7.1) is estimated by the rate ratio RR, i.e., the ratio of two observed attack rates:

$$\hat{VE} = 1 - RR$$
$$= 1 - \frac{AR_1}{AR_0}$$
$$= \frac{AR_0 - AR_1}{AR_0}$$

with AR_1 the observed attack rate among the subjects in the investigational vaccine group and AR_0 the observed attack rate among the control subjects. If the control is a placebo, then $\hat{VE}$ estimates the *absolute vaccine efficacy*, the proportion of infected cases prevented by the vaccine. If the control is an active vaccine, then $\hat{VE}$ estimates the *relative vaccine efficacy*, the relative decrease in number of infected cases.

Testing the null hypothesis that the absolute or relative vaccine efficacy is zero is equivalent to testing that the relative risk of infection is one. This null hypothesis can be tested with Pearson's chi-square test, or, if group sizes are small, the Suissa and Shuster exact test. If (LCL_θ, UCL_θ) is a two-sided $100(1 - \alpha)\%$ confidence interval for the relative risk θ then

$$(1 - UCL_\theta, 1 - LCL_\theta)$$

is a two-sided $100(1 - \alpha)\%$ confidence interval for the vaccine efficacy $VE = (1 - \theta)$.

Example 7.2. Blennow and colleagues report the result of a whole cell pertussis vaccine field efficacy study [50]. The study was a multi-centre trial, performed in Sweden, in the early nineteen-eighties. In this nonblinded trial, 525 infants aged 2 months who were born on days with an even number received three doses of vaccine one month apart, and 615 infants of the same age who were born on days with an odd number were enrolled as controls. The surveillance period, the age period between 6 and 23 months, was fixed. In the vaccinated group, 8 cases of pertussis occurred, compared to 47 in the control group. The estimated absolute vaccine efficacy was

$$\hat{VE} = 1 - \frac{8/525}{47/615} = 0.801.$$

In the vaccinated group, 80% of the expected cases of pertussis was prevented. The null hypothesis that the vaccine efficacy is zero could be rejected because the P-value for Pearson's chi-square test was <0.0001. The two-sided 95% Wilson-type confidence interval for the relative risk θ is $(0.097, 0.410)$, which corresponds to a 95% confidence interval for the vaccine efficacy VE of $(0.590, 0.903)$. Suppose that there had been the requirement of super-efficacy, with the requirement being that the vaccine efficacy exceeded 0.5. Because the lower bound of the confidence interval for vaccine efficacy falls above this bound, this requirement would have been met.

What often is overlooked is that the vaccine efficacy may depend on the length of the surveillance period, because for many vaccines the protection they afford wanes over time. In that case, the longer the length of the surveillance period, the lower the vaccine efficacy. A solution is to report the vaccine efficacy by sub-period, for example, first year after vaccination, second year after vaccination, etc.

7.4.2 Comparing Two Infection Rates

In trials in which the incidence of infection is measured by infection rates, the vaccine efficacy is estimated using the *infection rate ratio* as measure of the relative risk:

$$\hat{VE} = 1 - IRR,$$

with

$$IRR = \frac{s_1/T_1}{s_0/T_0}$$

the estimated infection rate ratio. Here, s_1 and T_1 and s_0 and T_0 are the number of infected cases and the total person-time in the investigational vaccine and the control group, respectively.

The interpretation of $\hat{VE}$ is similar to that of $\hat{VE}$ based on attack rates. The vaccine efficacy is the proportion of infected cases prevented by the vaccine during an arbitrary time unit if the control is a placebo, or the relative decrease in infected cases during an arbitrary time unit if the control is a vaccine.

The standard statistical test to compare two infection rates is a conditional exact test based on the conditional binomial distribution of the number of infected cases in the investigational vaccine group given the total number of cases in both groups. This test is based on the assumption that for both groups the number of cases is a realization from a Poisson distribution, with parameter λ_1 for the investigational vaccine group and λ_0 for the control group. Under the Poisson assumption, s_1 is binomially $B(s, \pi)$ distributed, conditional on s, the total number of cases, and with

$$\pi = \frac{T_1\lambda_1}{T_1\lambda_1 + T_0\lambda_0}. \tag{7.2}$$

The null hypothesis $H_0 : \lambda_1 = \lambda_0$ is tested by testing the equivalent null hypothesis $H_0 : \pi = T_1/(T_1 + T_0)$. Exact confidence limits for π are translated to exact confidence limits for $\theta = \lambda_1/\lambda_0$, the infection rate ratio, and for $VE = (1 - \theta)$, the vaccine efficacy. If $r = T_1/T_0$, the ratio of the total person-times, then

$$\theta = \frac{\pi}{r(1 - \pi)}.$$

Thus, if LCL_π and UCL_π are the exact lower and upper $100(1-\alpha)\%$ confidence limits for π (see Sect. 3.5.1), then

$$LCL_\theta = \frac{LCL_\pi}{r(1 - LCL_\pi)} \quad \text{and} \quad UCL_\theta = \frac{UCL_\pi}{r(1 - UCL_\pi)}$$

are exact $100(1-\alpha)\%$ confidence limits for θ, while

$$LCL_{VE} = 1 - UCL_\theta \quad \text{and} \quad UCL_{VE} = 1 - LCL_\theta$$

are exact lower and upper $100(1-\alpha)\%$ confidence limits for the vaccine efficacy VE.

Example 7.3. Urdaneta and colleagues report the results of the randomized, placebo-controlled, field efficacy trial of an SPf66 malaria vaccine [51]. The objective of the trial was to evaluate in nonimmune residents of a Brazilian endemic region, the efficacy of the SPf66 vaccine for all, as well as the first and second episodes of malaria infections separately for *P. falciparum* and *P. vivax*. A total of 800 subjects were enrolled in the trial. Of the initial cohort, 572 participants completed the vaccination schedule (3 doses), 287 in the vaccine group and 285 in the control group. In the vaccine group 76 first *P. falciparum* malaria episode occurred during a total follow-up of 12,178 person-weeks, compared to 85 episodes during a total follow-up of 11,698 person-weeks in the control group. The vaccine efficacy was estimated to be

$$1 - \frac{76/12{,}178}{85/11{,}698} = 0.141.$$

To test the null hypothesis that the vaccine efficacy is zero, the null hypothesis

$$H_0 : \pi = \frac{12{,}178}{12{,}178 + 11{,}698} = 0.510.$$

was tested. Under the null hypothesis the probability that the number of cases in the vaccine group is less or equal to 76 given that the total number of cases is 161 equals

$$\text{PROBBNML}(0.510, 161, 76) = 0.188.$$

The exact 95% two-sided confidence interval for π is $(0.393, 0.522)$. With $r = 12{,}178/11{,}698 = 1.041$, the following lower and upper limit of the 95% confidence interval for the infection rate ratio θ are obtained

$$LCL_\theta = \frac{0.393}{1.041(1 - 0.393)} = 0.622$$

and

$$UCL_\theta = \frac{0.552}{1.041(1 - 0.552)} = 1.184.$$

Thus, an exact 95% confidence interval for the vaccine efficacy VE is $(-0.184, 0.378)$. The trial did not yield evidence that the SPf66 vaccine protects against *P. falciparum* malaria infection.

7.4.3 Comparing Two Force of Infection Functions

A popular model for comparing two force of infection functions $f_1(t)$ and $f_0(t)$ is the Cox proportional hazards model. The model is based on the assumption that the functions are proportional, i.e., that their ratio is a constant:

$$f_1(t)/f_0(t) = \theta.$$

The constant θ is called the hazard rate ratio. A constant hazard rate ratio can be interpreted as follows: in case of a placebo-controlled field efficacy trial, the assumption is that in any (short) time-interval during the surveillance period the proportion of infected cases prevented is the same, independent of the number of infected cases in the control group. Or, to put it differently, that the vaccine efficacy is independent of the rate at which infected cases occur in the control group. This is a very plausible assumption (although one can think of a scenario with mostly mild cases during the first part of the surveillance period and mostly severe cases during the second part, in which case the proportional hazards assumption may not hold.) The proportional hazards assumption is a much weaker condition than the condition of a constant risk.

An attractive property of the Cox model is that the vaccine efficacy can be estimated without specifying $f_1(t)$ and $f_0(t)$. This is done with the procedure PHREG.

Example 7.4. Consider a hypothetical placebo-controlled field efficacy trial with a pandemic influenza vaccine. All subjects were vaccinated during a period of four weeks after the vaccine became available, and the length of the surveillance period varied per subject. Let t denote the time (days) to infection for infected subjects and the length of the surveillance period for noninfected subjects.

SAS Code 7.1 *Comparing Two Groups of Times to Infection under the Proportional Hazards Assumption*

```
data;
    input group t infected;
datalines;
0   30 1
0  113 0
.
.
```

```
1  68 1
1  70 0
run;

proc phreg;
    model t*infected(0)=group / risklimits;
run;
```

Here, `group=1` for the investigational vaccine group and `group=0` for the control group, `t` is the time to infection (if `infected=1`) or the length of the surveillance period for noninfected subjects (if `infected=0`). In the model statement, the value for censored observations (here: 0) has to be specified.

SAS Output 7.1

```
                      The PHREG Procedure
              Analysis of Maximum Likelihood Estimates

            Parameter Standard   Chi-                Hazard   95% Hazard Ratio
Variable DF  Estimate    Error Square Pr > ChiSq   Ratio   Confidence Limits

group    1   -1.55782  0.55288 7.9391     0.0048   0.211   0.071        0.622
```

The parameter being estimated is $\log(\theta)$. The estimated hazard rate ratio is 0.211, with 95% confidence interval $(0.071, 0.622)$. Thus, the estimated vaccine efficacy is $\hat{VE} = 0.789$, with 95% confidence interval $(0.378, 0.929)$.

If the proportional hazards ratio assumption holds, then (7.2) implies

$$\theta = \frac{\log_e[1 - \pi_1(t)]}{\log_e[1 - \pi_0(t)]}. \tag{7.3}$$

The Taylor series for $-\log_e(1 - \pi)$ is

$$-\log_e(1 - \pi) = \pi + \frac{\pi^2}{2!} + \frac{\pi^3}{3!} + \cdots$$

Thus, for small values for π_1 and π_0

$$\theta \approx \frac{\pi_1(t)}{\pi_0(t)}.$$

Meaning that the estimated vaccine efficacy

$$VE \approx 1 - \frac{\pi_1(t)}{\pi_0(t)}.$$

7.5 Recurrent Infections

A *recurrent infection* is an infection caused by the same pathogen in a subject who has experienced at least one infection before. The statistical analysis of recurrent infection data is extremely complex. There are two major methodological challenges to address. The first challenge is the possibility that the risk of a next infection is affected by previous infections. It could, for example, be the case that a first infection makes a subject more susceptible for infection. The second challenge is that the risk of infection may be different among subjects. In that case, it can be that the more prone subjects are contributing more infections than less prone subjects. With one exception, for neither challenge a simple statistical solution exists. But there are other challenges as well. An example of an infectious disease with recurrent episodes in patients is malaria. A group of experts on this disease reported that it is difficult to define when a malaria episode has ended and when following treatment a subject becomes susceptible to a further episode [52]. For this reason and the statistical complexity involved in analysing recurrent episodes, the consensus of the group was that the primary endpoint in a malaria trial should be time to first episode of clinical malaria, although the members admitted that the total number of episodes of malaria in trial participants might better measure the total burden of malaria in the community. This is also pointed out by Janh-Eimermacher, du Prel and Schmitt, who give an excellent overview of the statistical methods used to assess vaccine efficacy for the prevention of acute otitis media (AOM) by pneumococcal vaccination [53]. They note that: 'The proportion of subjects with at least one episode might be equal in two vaccine groups, whereas the groups differ substantially in the total number of episodes. (...) Decreasing the total number of AOM might reduce the total costs for treating AOM, improve the problem of antibiotic resistance caused by broad antibiotic use in treatment of AOM and reduce the long-term effects cause by recurrent episodes.

The drawback of considering only the first infection is that not all available information is used, which can lead to an under- or overestimation of the value of the benefit of a vaccine.

7.5.1 Average Number of Episodes Experienced by a Subject

For the average number of infectious episodes experienced by a subject to be a meaningful concept, it must be linked to a fixed time interval. The time interval can be either a fixed time span, say, a year, or a fixed time period, the first two years after vaccination or a specified age period, for example. The time interval must be fixed, but the surveillance period need not necessarily to be of fixed length. This depends on whether the risk of a next episode is unaffected by previous episodes. If this is the case then, if the interest is in the average number of episodes experienced by a child during the first five years of live, the only restriction is that the surveillance period should fall within this age period. If, however, the risk of a next episode is affected by previous episodes, then length of the surveillance period must be fixed.

The statistical analysis of intra-individual numbers of episodes depends on which of the following two conditions are met:

1. The risk of a next episode is unaffected by previous episodes
2. The risk of infection does not differ between subjects

If both conditions are met, then the statistical analysis is straightforward. In that case, the average number of episodes experienced by a subject during, say, a year, can be estimated by the total number of episodes (i.e., the sum of the intra-individual numbers of episodes) divided by the total number of person-years. The total number of episodes will be a realization from a Poisson distribution, and to compare the total number of episodes between vaccine groups the methods in Sect. 7.4.2 can be applied.

In practice, it will be unlikely that the second condition is met. More likely is that the risk of infection differs between subjects. This is called overdispersion. If overdispersion is ignored the variance of the estimator is underestimated. Let the random variable $\mathbf{Y}$ be the intra-individual number of episodes. If the Poisson model would hold, then the variance of $\mathbf{Y}$ would be equal to its expectation:

$$\text{var}(\mathbf{Y}) = E(\mathbf{Y}) = \mu.$$

A much applied approach to deal with count data that exhibit overdispersion is to assume that the individual risk itself may be regarded as a random variable. In that case, the probability distribution will be a compound (mixed) distribution with

$$\text{var}(\mathbf{Y}) = V(\mu) > \mu.$$

There are two standard choices for the variance function $V(\mu)$. The first is to assume that the individual risks are random draws from a gamma distribution. The resulting compound distribution is the negative binomial distribution, also known as the gamma-Poisson (mixture) distribution. This leads to a quadratic variance function:

$$V(\mu) = \mu + \kappa\mu^2, \quad \kappa > 0.$$

The second standard choice for the variance function is

$$V(\mu) = \sqrt{\zeta}\mu, \quad \zeta > 1.$$

This is a convenient choice, which does not require a specification of the compound distribution. The parameters ζ and κ are called the dispersion parameters [54].

Both variance models can be fitted with the SAS procedure GENMOD. The GENMOD procedure fits a generalized linear model to the data by maximum likelihood estimation. The standard link function for (overdispersed) count data

is the log function. The SAS code to fit the an overdispersed Poisson model with
$V(\mu) = \sqrt{\zeta}\mu$ to count data is

SAS Code 7.2 *Fitting an Overdispersed Poisson Model to Intra-Individual Number of Episodes*

```
data;
   input vaccine number_of_episodes
         length_surveillance_period;
   loglsp=log(length_surveillance_period);
datalines;
0 0 120
0 1 100
0 1 120
0 3  90
0 3 130
0 4 150
0 5 100
1 0 120
1 0 100
1 1  90
1 2 120
1 3  80
1 4 130
run;

proc genmod;
   class vaccine;
   model number_of_episodes=vaccine /
      dist=poisson link=log scale=deviance
   offset=loglsp type3;
   estimate "infection rate ratio" vaccine -1 1;
run;
```

SAS Output 7.2

```
               Model Information

   Distribution                   Poisson
   Link Function                      Log
   Dependent Variable    number_of_episodes
   Offset Variable                 loglsp

   Criteria For Assessing Goodness Of Fit

   Criterion         DF    Value   Value/DF
   Deviance          11  20.8621     1.8966
   Pearson Chi-Square 11  16.8524     1.5320
```

```
Analysis Of Maximum Likelihood Parameter Estimates

                         Standard  Wald 95% Confidence       Wald
Parameter   DF  Estimate   Error        Limits         Chi-Square  Pr > ChiSq
Intercept    1  -4.1589   0.4355   -5.0124   -3.3053       91.20     <.0001
vaccine   0  1   0.2951   0.5488   -0.7806    1.3708        0.29     0.5908
vaccine   1  0   0.0000   0.0000    0.0000    0.0000         .          .
Scale        0   1.3772   0.0000    1.3772    1.3772

             Contrast Estimate Results

                         Mean          Mean
Label                  Estimate  Confidence Limits
infection rate ratio    0.7445   0.2539     2.1828
```

There are two criteria for overdispersion, the `Deviance` divided by its degrees of freedom and the `Pearson chi-square` divided by its degrees of freedom, 1.8966 and 1.5320, respectively. Both are greater than 1.0, which indicate overdispersion. Both statistics can be used as estimates of the dispersion parameter. To do this, either

$$\text{scale=Pearson} \quad \text{or} \quad \text{scale=deviance}$$

must be inserted as an option in the model statement, to obtain an overdispersed Poisson distribution. First, estimates are obtained for the nonoverdispersed Poisson model. Next, the scale parameter is estimated by either the square root of the Pearson statistic or the square root of the deviance statistics, and the standard errors are adjusted by multiplying them by the value for the scale statistic

$$1.3772 = \sqrt{1.8966},$$

making the statistical tests more conservative. With the `estimate` statement a point and an interval estimate of the infection rate ratio is obtained. For the hypothetical data, an estimate of the vaccine efficacy is thus

$$\hat{VE} = 1 - 0.7445 = 0.2555.$$

To fit a negative binomial distribution to the data the option `dist=poisson` should be replaced with `dist=negbin`, and the `scale` option should be deleted:

SAS Code 7.3 *Fitting a Negative Binomial Model to Intra-Individual Number of Episodes*

```
proc genmod;
   class vaccine;
   model noe=vaccine / dist=negbin link=log offset=loglsp type3;
   estimate "infection rate ratio" vaccine -1 1;
run;
```

SAS Output 7.3

```
Analysis Of Maximum Likelihood Parameter Estimates

                        Standard  Wald 95% Confidence       Wald
Parameter    DF  Estimate   Error        Limits         Chi-Square  Pr > ChiSq

Intercept     1   -4.1565  0.3624  -4.8667    -3.4463      131.57     <.0001
vaccine      0 1   0.3000  0.4657  -0.6127     1.2127        0.42     0.5194
vaccine      1 0   0.0000  0.0000   0.0000     0.0000
Dispersion    1   0.1831  0.3260  -0.2000     0.8220

          Contrast Estimate Results

                          Mean          Mean
Label                   Estimate   Confidence Limits
infection rate ratio     0.7408     0.2974     1.8453
```

7.5.1.1 Time to a Next Episode

Andersen and Gill have proposed generalized Cox proportional hazard method that takes recurrent episodes into account [55]. As with the Cox proportional hazard model, the generalization is based on the assumption that the hazard rate is proportional between both groups over time. The method requires that the assumption that the risk of a next episode is unaffected by previous episodes, and if this assumption is not met misleading results may be obtained.

7.6 Sample Size Estimation

7.6.1 Trials Comparing Two Attack Rates

To estimate the required sample size of a vaccine field efficacy trial comparing two attacks rates, three parameters need to be specified: the expected attack rate in the control group, the expected vaccine efficacy and the criterion for vaccine efficacy. With the vaccine efficacy specified, the expected attack rate in the investigational vaccine group can be calculated. If the criterion for vaccine efficacy is that the efficacy must be greater than zero, then the required sample size can be computed with SAS code 3.2.

If super efficacy needs to be demonstrated, more complicated computations are required. Having to demonstrating that the vaccine efficacy VE is $> \delta$ is equivalent to having to demonstrate that the relative risk θ for the investigational versus the placebo group is $< (1 - \delta)$. With the procedure POWER, the required sample size for testing $H_0 : \theta \geq (1 - \delta)$ against the one-sided alternative $H_1 : \theta < (1 - \delta)$ can be approximated.

Example 7.5. Consider a placebo-controlled influenza vaccine field efficacy trial. Assume that the expected attack rate in the control group is 0.25 and that the expected vaccine efficacy is 0.8. An expected vaccine efficacy of 0.8 corresponds to an expected attack rate of

$$0.25 \times (1 - 0.8) = 0.05$$

in the vaccine group. If the null hypothesis is that the vaccine efficacy is zero, then SAS code 3.2 computes the sample size to be $2 \times 65 = 130$ to be secured of a power of 0.9 (for a two-sided significance level of 0.05). The FDA/CBER criterion for super efficacy for influenza vaccines is 0.4. This means that to demonstrate super efficacy the null hypothesis $H_0 : \theta \geq 0.6$ must be tested against the alternative hypothesis $H_1 : \theta < 0.6$, at the one-sided significance level 0.025. An expected vaccine efficacy of 0.8 corresponds to an expected relative risk of $(1 - 0.8) = 0.2$. A first attempt might be to approximate the required sample size using either one of the following two SAS codes:

SAS Codes 7.4 *Approximate Sample Size Estimation for Demonstrating Super Efficacy when Comparing Two Attack Rates*

```
proc power;
   twosamplefreq test=pchi
   alpha=0.025 sides=1
   groupproportions=(0.25,0.05)
   nullpdiff=-0.10   /* 0.25(1-0.4) - 0.25 */
   power=0.9 npergroup=.
run;

proc power;
  twosamplefreq test=pchi
   alpha=0.025 sides=1
   refproportion=0.25
   relativerisk=0.2
   nullrelativerisk=0.6
   power=0.9  npergroup=.
run;
```

When either one of these codes are run, the computed sample size is found to be $2 \times 260 = 520$ subjects. A warning, however, is at its place here. The power computed here is the power for rejection the one-sided null hypothesis given above using Pearson's chi-square statistic with the standard error in (3.5). Monte Carlo simulation results suggest that the sample size estimate is correct for trials with the rate difference as outcome, but that they are conservative for trials with the rate ratio as outcome and with a Wilson-type confidence interval for the relative risk (see Sect. 3.5.2). Monte Carlo simulations estimate the latter power to be > 0.99, and that to be secured of a power of 0.9, a sample size of only $2 \times 140 = 280$ (!) would be required.

7.6.2 Trials Comparing Two Infection Rates

The procedure to determine the sample size for a vaccine field efficacy trial in which two infection rates are to be compared is as follows. Let s_0 denote the expected number of placebo cases, s_1 the expected number of cases in the investigational vaccine group and $s = (s_0 + s_1)$ the expected total number of cases. For s fixed, determine the largest number for s_1 such that the lower limit of the two-sided $100(1-\alpha)\%$ confidence interval for the vaccine efficacy VE is > 0, or $> \delta$ if super efficacy must be demonstrated. The actual power of the trial is then the probability that the number of cases in the investigational vaccine group is less or equal to s_1. Find the value for s such that the actual power is equal to the desired power.

The formula for the expected total number of cases is

$$
\begin{aligned}
s &= n_0\lambda_0 + n_1\lambda_1 \\
&= n_0\lambda_0 + n_1\lambda_0(1 - VE) \\
&= n_0\lambda_0 + n_0 r\lambda_0(1 - VE) \\
&= n_0\lambda_0[1 + r(1 - VE)],
\end{aligned}
$$

where λ_0 and λ_1 are the expected infection rates during the length of the planned average surveillance period, and $r = n_1/n_0$, the randomization ratio. Hence,

$$
n_0 = \frac{s}{\lambda_0[1 + r(1 - VE)]}.
$$

Example 7.6. An investigator wants to estimate the sample size for a vaccine field efficacy study with a planned average surveillance period of 24 months, a 2:1 randomization ratio, an expected infection rate in the placebo group of 0.05 cases per 24 months, an expected vaccine efficacy of 0.7 and with $\delta = 0.4$ as bound for super efficacy. The sample size can be computed with SAS code 7.5. If the desired power is 0.9, then the required number of investigational cases is 40 and the required number of total cases 92. This corresponds to a sample size of 3,453 subjects, 2,302 to be randomized to the investigational vaccine and 1,151 to placebo.

SAS Code 7.5 *Sample Size Computation for Comparing Two Infection Rates*

```
data;
    ve=0.7;          /* expected vaccine efficacy */
    delta=0.4;       /* bound for super efficacy  */
    lambda0=0.05;    /* expected infection rate control group */
    r=2;             /* randomization ratio n1/n0     */
    alpha=0.05;      /* two-sided significance level */
    power=0.9;       /* desired statistical power     */

    pi=r*(1-ve)/(r*(1-ve)+1);
    s=1; actual_power=.;
```

```
do while (actual_power lt power);
    s=s+1; s1=s; ready=0;
    do while (not ready);
        s1=s1-1; s0=s-s1;
        fu=finv(1-alpha/2,2*(s1+1),2*s0);
        ucl_pi=(s1+1)*fu/(s0+(s1+1)*fu);  /* upper CL for pi     */
        ucl_theta=ucl_pi/(r*(1-ucl_pi));  /* upper CL for theta */
        lcl_ve=1-ucl_theta;               /* lower CL for VE    */
        ready=((lcl_ve>delta)|(s1=0));    /* largest c1 for which lower  */
    end;                                  /* CL for vaccine efficacy>bound */
    actual_power=probbnml(pi,s,s1);       /* probability of c1 or less   */
    end;                                  /* cases in experimental group */
    n0=int(s/(lambda0*(1+r*(1-ve))))+1;
    n1=r*n0;
    output;
run;
```

7.6.3   Trials Comparing Two Forces of Infection

Power and sample size calculations for comparing two forces of infection can be
performed with the `twosamplesurvival` statement of the procedure POWER.
If a proportional hazards model is assumed, the `test` option should be set to
`test=logrank`. Sample size estimation for comparing two forces of infection
requires detailed specification of the design of the study. The reader is therefore
referred to the chapter on the POWER procedure in the SAS/STAT User's Guide for
a description of the different options to specify the details of the design of the study.

Chapter 8
Correlates of Protection

8.1 Introduction

An immune *correlate of protection* is an immunological assay (either humoral or cellular) that predicts protection against infection or disease. Correlates of protection are of great importance because they can be used as surrogate endpoints for vaccine efficacy in clinical vaccine trials. Immunological vaccine trials are much less costly and much less time-consuming than field efficacy trials. Correlates of protection constitute the scientific basis for improving existing vaccines and introducing new ones. Under certain conditions, they can be used to predict vaccine efficacy in populations other than the one in which efficacy was demonstrated. For these reasons, finding correlates of protection is one of the 14 Grand Challenges of Global Health of the National Institutes of Health (NIH) and the Bill and Melinda Gates Foundation.

For many infectious diseases, it has been established that particular antibody assays are correlates of protection. Infectious diseases for which this has been demonstrated that are, amongst others, influenza, hepatitis A and hepatitis B, meningococcal disease, tetanus, measles, mumps, rubella and polio. Recently, Forrest and co-workers showed that the interferon-γ ELISPOT assay is a correlate of protection for influenza in young children [56].

In the scientific literature, the term correlate of protection is used in two different meanings. First, it is used in the meaning of an immunological assay that predicts protection against infection or disease. Second, it is used in the meaning of a clear-cut value for, say, an antibody assay, above which subjects are protected. In that case, often the protective level itself is called a correlate of protection. Siber discusses several examples of this [57]. One example he discusses is diphtheria, an upper respiratory tract illness. The disease is caused by the bacterium *Corynebacterium diphtheria*. A characteristic symptom of the disease is a swollen neck, sometimes referred to as 'bull neck'. During an outbreak of the disease in Copenhagen in 1943–1944 sera of 425 patients were collected. In 95% of the cases, diphtheria antitoxin levels were greater than 0.01 antitoxin units. From these observations, it was concluded that this antitoxin level is a correlate of protection. Here,

J. Nauta, *Statistics in Clinical Vaccine Trials*, DOI 10.1007/978-3-642-14691-6_8,
© Springer-Verlag Berlin Heidelberg 2011

correlate of protection will be used in the first meaning only. For a protective assay level, the term threshold of protection is used.

The correlation between an immunological assay and protection from infection can be assessed in vaccine field efficacy trials, challenge studies, household transmission studies and observational cohort studies.

8.2   The Protection Curve

A *protection curve* is a mathematical function specifying the relationship between log-transformed immunogenicity values and the probability of protection against infection, *conditional* on exposure to the pathogen. For convenience, here it will be assumed that the immunogenicity values are antibody titres.

An obvious choice for the protection curve is the sigmoid logistic function:

$$pc(\log t) = \frac{1}{1 + \exp(\alpha + \beta \log t)},$$

with $\alpha > 0$, $\beta < 0$ and t the antibody titre value. The standard logit function has two parameters, α and β, to be estimated from the data. The intercept α is the location parameter of the protection curve, and reflects the protection not mediated by antibodies. The slope β contributes to the steepness of the curve, the larger $|\beta|$, the steeper the curve. An antibody assay is a correlate of protection if $\beta \neq 0$.

8.3   Estimating the Protection Curve

8.3.1   Estimating the Protection Curve from Challenge Study Data

A *challenge study* is a study in which vaccinated volunteers are challenged with a pathogen. Challenge studies are an important tool in clinical vaccine development, because they can furnish proof-of-concept for an experimental vaccine and accelerate progress towards phase III trials. Imperial College in London is famous for its malaria challenge studies. A group of volunteers is vaccinated with the experimental vaccine, while a second group of volunteers serves as control group. At a predefined number of days after the (final) vaccination, the subjects are infected with malaria. Five infected mosquitoes wait under a mesh draped over a paper coffee cup. The volunteer rests his arm over the cup to allow the mosquitoes to bite. Monitoring takes place twice daily. Subjects are treated with anti-malarial drug chloroquine (to prevents the development of malaria parasites in the blood) after the first confirmed positive blood film or at day 21 if no parasitemia was detected. Protection can be complete or partial. Complete protection is where none of the vaccinated volunteers do develop malaria (but all unvaccinated control volunteers do). Partial protection is where there is a delay in the onset of malaria in the vaccinated volunteers, meaning that the immune system is controlling the infection to start but is ultimately overwhelmed.

Challenge studies have a significant limitation, namely that it is often not possible to expose the volunteers to a wild-type strain, i.e., a strain found in nature, because this would be too dangerous. In that case, the volunteers are challenged with a laboratory-adapted strain, which is a strain weakened by passing. The limitation then is the fidelity of a laboratory-adapted challenge model to natural infection.

Example 8.1. Hobson and colleagues studied the role of haemagglutination inhibition in protection against challenge infection with influenza viruses [58]. Four-hundred-and-sixty-two adult volunteers (industrial workers) who were randomly assigned to be vaccinated with a live or a killed influenza vaccine or placebo, whilst others were left unvaccinated. Two to three weeks later they were challenged, by means of intranasal inoculation with a living, partially attenuated strains of an influenza B virus. Serum samples for anti-HA antibody determination by means of the HI test were drawn immediately before virus challenge. Nasal swabs were taken for virus isolation studies wherever possible 48 h after challenge. In total, 135 of the volunteers got infected and 327 remained infection free. In Fig. 8.1, the observed proportions of protected subjects is shown. A logit protection curve was fitted to the challenge data. In SAS, there are several procedures that can be used to fit a logit protection curve to challenge data. The most flexible of these is procedure NLMIXED.

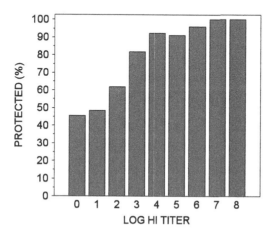

Fig. 8.1 Observed protection in an influenza vaccine challenge study

SAS Code 8.1 *Fitting a Logistic Protection Curve to Challenge Study Data*

```
proc nlmixed;
   parms alpha=0.1 beta=-0.1;
   eta=exp(alpha+beta*logtitre);
   p=1/(1+eta);
   model protected ~ binomial(1,p);
   predict alpha+beta*logtitre out=fitted;
run;

data pcurve; set fitted;
   pcurve=1/(1+exp(pred));
   uclpcurve=1/(1+exp(lower));
   lclpcurve=1/(1+exp(upper));
run;
```

The variable `logtitre` should contain the log transformed antibody titres. The outcome variable `protected` must be a binary variable, being set to 1 for subjects who were protected after challenge and to 0 for subjects who got infected. The variable `pcurve` will contain the values for the fitted protection curve, and the variables `lclpcurve` and `uclpcurve` will contain the lower and upper limit of a two-sided 95% confidence interval for $pc(\log t)$. Together, these confidence intervals constitute a 95% point-wise confidence band for the protection curve. The fitted protection curve is displayed in Fig. 8.2.

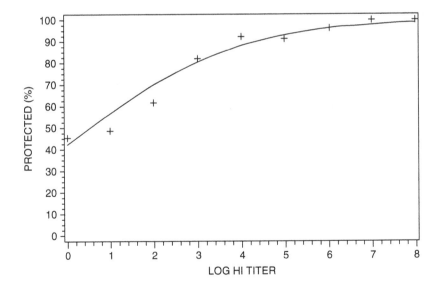

Fig. 8.2 Protection curve fitted to the challenge study data in Fig. 8.1

SAS Output 8.1

```
                          Parameter Estimates

                   Standard
Parameter  Estimate   Error    DF  t Value  Pr > |t|  Alpha    Lower     Upper

alpha       0.3130   0.1567    462    2.00    0.0463   0.05   0.005164   0.6209
beta       -0.5794   0.06680   462   -8.67    <.0001   0.05  -0.7107    -0.4481
```

8.3.2 Estimating a Protection Curve from Vaccine Field Efficacy Study Data

In this section, it will be explained how a protection curve can be estimated from vaccine field efficacy data. A major difference between a challenge and a field efficacy study is that in a challenge study all subjects are exposed to the pathogen, while in a field efficacy study only a fraction of the subjects is exposed. This has to be allowed for in the model fitted to the data. This is done by applying a simple rule for conditional probabilities.

Consider a vaccine field efficacy data with a fixed surveillance period and with a particular antibody titre measured at a defined time point after vaccination. The probability of not getting infected during the surveillance period is

$$Pr(\text{not Infected}) = Pr(\text{not Infected} \mid \text{not Exposed})\, Pr(\text{not Exposed})$$
$$+$$
$$Pr(\text{Protected} \mid \text{Exposed})\, Pr(\text{Exposed}).$$

Because a subject cannot get infected if not exposed,

$$Pr(\text{not Infected} \mid \text{not Exposed}) = 1.$$

And because

$$Pr(\text{not Exposed}) = 1 - Pr(\text{Exposed}),$$

the above equation can be rewritten as

$$Pr(\text{not Infected}) = 1 - Pr(\text{Exposed})$$
$$+$$
$$Pr(\text{Exposed})\, Pr(\text{Protected} \mid \text{Exposed}).$$

It is the probability

$$Pr(\text{Protected} \mid \text{Exposed})$$

that the interest is in, because this is the probability being modelled by the protection function. The expression above gives the model to be fitted: the protection curve plus Pr(Exposed).

Let P_E denote the probability that a subject is exposed to the pathogen, and let $pc(\log t)$ denote the protection curve, with t the antibody titre value. Then

$$\text{Pr(not Infected} \mid t) = (1 - P_E) + P_E \ pc(\log t).$$

If it is further assumed the $pc(\log t)$ is the logistic function, then

$$\text{Pr(not Infected} \mid t) = (1 - P_E) + \frac{P_E}{1 + \exp(\alpha + \beta \log t)}.$$

This model is akin to the scaled logistic function proposed by Dunning [59]. Because the model separately parameterizes exposure, the protection curve can be estimated. This model also can be fitted with procedure NLMIXED.

Example 8.2. White and co-workers report the result of field efficacy trials with a live attenuated varicella (chickenpox) vaccine conducted between 1987 and 1989 [60]. Four thousand forty-two healthy children and adolescents, ages 12 months to 17 years, were vaccinated with a single dose of the vaccine. During the first and second years of follow-up, 2.1 and 2.4% of the vaccinees developed varicella. In Fig. 8.3, the incidence of varicella is shown by the 6-week postvaccination glycoprotein-based (gp) ELISA assay titre. The following SAS code can be used to fit a protection curve to the field efficacy data in Fig. 8.3:

SAS Code 8.2 *Estimating a Logistic Protection Curve from Field Efficacy Study Data*

```
proc nlmixed;
    parms pe=0.1 alpha=1 beta=-1;
    eta=exp(alpha+beta*logtitre);
    p=(1-pe)+pe/(1+eta);
    model notinfected ~ binomial(1,p);
    predict alpha+beta*logtitre out=fitted;
run;

data pcurve; set fitted;
    pcurve=1/(1+exp(pred));
    uclpcurve=1/(1+exp(lower));
    lclpcurve=1/(1+exp(upper));
run;
```

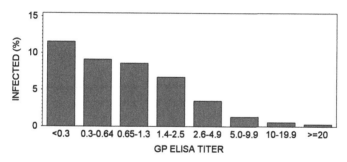

Fig. 8.3 Incidence of varicella in a field efficacy study

SAS Output 8.2

				Parameter Estimates				
Parameter	Estimate	Standard Error	DF	t Value	Pr > |t|	Alpha	Lower	Upper
pe	0.1156	0.02561	3459	4.51	<.0001	0.05	0.06538	0.1658
alpha	1.0182	0.7377	3459	1.38	0.1676	0.05	-0.4282	2.4647
beta	-1.4812	0.2733	3459	-5.42	<.0001	0.05	-2.0171	-0.9453

The proportion of trial participants in the trial that was exposed to the varicella zoster virus is estimated to be 0.116.

8.3.3 Predicting Vaccine Efficacy

An estimated protection curve can be used to predict vaccine efficacy:

$$VE_{\text{predicted}} = 1 - \frac{\sum_{i=1}^{n_1} pc(\log t_{1i})/n_1}{\sum_{i=1}^{n_0} pc(\log t_{0i})/n_0},$$

where t_{1i} is the antibody value of the ith subject in the investigational vaccine group and t_{oi} the antibody value of the ith subject in the placebo group. This requires the assumption that the protection curve for placebo subjects is the same curve as that for vaccinated subjects. This need not to be so, however. In that case, the protection curve for placebo subjects should be estimated separately.

8.4 Threshold of Protection

Underlying the concept of a protection threshold is the idea that there is an assay value above which most subjects are protected from infection, and below which most subjects are not. This would require a very steep protection curve, the so-called step curve. The examples of protection curves fitted in the previous section show that this is often not the case and that the curve will be nonsteep. In that case, an obvious definition of a protection threshold is the antibody value T_P for which the predicted probability of protection is 0.5:

$$pc(\log T_P) = 0.5.$$

If the fitted protection curve is the logistic one, then it is easy to see that whether the antibody titres are $\log_e$ transformed, the estimate of T_P is the exponential of minus the ratio of the fitted intercept and slope:

$$\hat{T}_P = e^{-\hat{\alpha}/\hat{\beta}}.$$

A confidence interval for T_P can be obtained by using in NLMIXED the following `predict` statement:

```
predict -alpha/beta;
```

The antilogs of the values for `lower` and `upper` then constitute a two-sided confidence interval for T_P.

Example 8.2. (continued) The estimate for the threshold of protection for varicella is a 6-week gp ELISA titre of

$$e^{1.0182/1.4812} = 2.0,$$

with as 95% confidence interval (0.9, 4.3).

An alternative method to estimate a threshold of protection is the *Chang and Kohberger method* [61]. To find a threshold of protection T'_P, the following equation is solved:

$$\frac{Pr(t < T'_P \mid \text{Vaccinated})}{Pr(t < T'_P \mid \text{not Vaccinated})} = \frac{Pr(\text{Infected} \mid \text{Vaccinated})}{Pr(\text{Infected} \mid \text{not Vaccinated})}$$

Chang and Kohberger applied their method to aggregate field efficacy data of trials with pneumococcal conjugate vaccine formulations, and found a serotype 19F IgG antibody threshold of 0.4 μg/ml. The method requires (a) that the vaccine efficacy is known, and (b) that the antibody levels in the control group show substantial variability. In practice, only an estimate $\hat{VE}$ of the vaccine efficacy will be available, in which case the equation to solve becomes

$$\frac{Pr(t < T'_P \mid \text{Vaccinated})}{Pr(t < T'_P \mid \text{not Vaccinated})} = 1 - \hat{VE}.$$

If the control is a placebo and the post-vaccination antibody levels in this group are all undetectable or very low, then

$$\Pr(t < T'_P | \text{not Vaccinated}) = 1.$$

In that case, solving the equation above will not lead to a sensible result. But even if (b) is met, care should be taken. The Chang and Kohberger method is based on the assumption that the threshold T'_P is the same for both the investigational and the control group. This need not be the case. If the control is a placebo, the antibody levels in the control group will be due to responding to natural infection, while the antibody levels in the investigational group will be due to responding to artificial infection. These antibody responses can be qualitatively different. This may also be the case if the control vaccine is a totally different type of vaccine (e.g., intranasal) than the investigational vaccine.

8.5 Discussion

The protection curve models the probability of protection as a function of the antibody titre. There are some subtleties to keep in mind, though, when estimating or interpreting a protection curve. First, if antibody levels do not decline over time, then the relationship between the antibody titres and protection is time-independent, i.e., not dependent on the timing of the blood sampling for antibody determination. This is of importance for studying pathogenesis, which requires that the probability of protection is related to the antibody level at the time of exposure. If antibody levels, however, do decline over time, then the relationship will be time-dependent. If the antibody levels in the model were measured three weeks after the vaccination, then the model may not be valid for antibody levels measured at later time points. It is therefore of importance to always state clearly the timing of the blood sampling (e.g., six week after vaccination, as in Example 8.2). If the time point at which the antibody titre is measured is the same for all trial participants, which is usually the case, then the (strength of the) relationship will depend on whether this point was before, at or after the time of the peak levels. The probability of protection will depend on the antibody level at the time of exposure. The lower the correlation between the measured antibody levels and the levels at the time of exposure the weaker the relationship will be. Second, the relationship may be dependent upon the length of the surveillance period, either because antibodies decline during the surveillance period.

An important but difficult to answer question is the generalizability of estimated protection curves. For example, how generalizible are protection curves estimated from challenge data? This will depend, amongst others, on how similar the challenge strain is to the wild-type strain and on the volunteers, usually healthy young male and female students. Can a protection curve estimated from data collected in adults be assumed to be universal, e.g., being applicable to an elderly population as well? Could the threshold be serotype dependent, or may it be different for some serotypes? Does it perhaps matter if antibodies are naturally or artificially acquired? The traditional threshold of protection for seasonal influenza vaccines, a haemagglutination inhibition titre of ≥ 40, is now being used for pandemic influenza vaccines as well, although there is little scientific justification for this.

Chapter 9
Safety of Vaccines

9.1 Ensuring Vaccine Safety

To proof the safety of a vaccine is much more challenging than proving its efficacy. Many vaccines are administered to several hundred million, often healthy people (e.g., childhood vaccines), in which case even extremely rare but serious adverse vaccine events can come to light, which may change the opinion of the medical community on the benefit/risk ratio. If a rare but serious condition occurs in, say, 0.1% of the target population and a vaccine doubles the risk to 0.2%, then there will be an additional 1,000 cases for every million persons vaccinated. A recent example of such an increased risk was a combination MMRV(measles, mumps, rubella, varicella) vaccine for children aged twelve months through twelve years, as alternative for two separate MMR and V vaccines. Post-licensure surveillance by the Vaccine Safety Datalink, a resource established by the United States Centers for Disease Control and Prevention (CDC) to investigate safety hypotheses using administrative databases of health maintenance organizations, detected a signal for increased febrile seizures in children between one and two years of age who had received the MMRV vaccine compared with those who had received the MMR vaccine. A febrile seizure, also known as fever fit or a fever convulsion, may happen with any condition that causes a sudden change in body temperature. These seizures can be caused by common childhood illnesses such as ear infection. During a febrile seizure, a child often has spasms and may lose consciousness. Vaccination may cause the body temperature to rise. It has been estimated that children who receive the combination MMRV vaccine are twice as likely to have a febrile seizure seven to ten days after the vaccination than children who get separate MMR and V vaccines. Because of this increased risk, MMRV is no longer advised over MMR + V separately. To assess the causal link between a vaccine and a serious condition from observational data is extremely difficult. Hepatitis B vaccination has been linked to rheumatoid arthritis, lupus erythematosus, diabetes mellitus, acute leukaemia, chronic fatigue syndrome and hair loss, but none of this has been proven conclusively.

Another recent example is the emotionally charged thiomersal controversy. Thiomersal is an ethyl-mercury-containing preservative, which has been used to prevent bacterial contamination of vaccines since the 1930s. In 1999, the United

J. Nauta, *Statistics in Clinical Vaccine Trials*, DOI 10.1007/978-3-642-14691-6_9,
© Springer-Verlag Berlin Heidelberg 2011

States Food and Drug Administration (FDA) noticed that with the then vaccination program for children, infants, by the age of 6 months, could have received a total of 187.5 microgram of mercury. This lead to concerns with the CDC and the American Academy of Pediatrics (AAP), the two organizations responsible for making childhood vaccine recommendations. Vaccine manufacturers were asked to remove thiomersal from their vaccines. This recommendation confused both parents and health care workers about the safety of vaccines. Studies were performed to investigate whether thiomersal in vaccines caused neuro-developmental or psychological problems. Evidence could not be found. Despite this, in 2000, the notion that thiomerosal can cause autism emerged. This notion was disproved by several epidemiologic studies. The controversy has led to considerable medical and social damage.

Pre-licensure clinical vaccine trials typically focus on a special class of adverse events known as local and systemic reactions, on abnormal laboratory values and, depending on the vaccine, on abnormal vital signs values (body temperature and blood pressure), and on none-rare other adverse events. Phase I is mostly of an exploratory nature, to demonstrate initial safety, and often there is no statistical inference. Phase II is to quantify the occurrence of local and systemic reactions and laboratory abnormalities. Phase III is to evaluate less but nonrare common adverse events. Post-licensure the focus moves to rare but serious events by means of surveillance.

9.2   Vaccine Safety Surveillance

Vaccine manufacturers are required to report to the registration authorities all serious adverse events of which they become aware. Post-licensure (post-marketing) *vaccine safety surveillance* further relies on physicians and others to voluntarily submit reports of illness after vaccination. This is both the strength and the weakness of surveillance. The strength is that the system has proven to be able to detect very rare but serious risks of specific vaccinations. The weakness is that the reporting system has considerable limitations, including variability in the quality if the reports, biased reporting and underreporting, inadequate denominator data, absence of unvaccinated controls groups and the inability to determine whether a vaccine caused the adverse event in any individual report.

The problems vaccine surveillance is faced with are tremendous. Ellenberg gives a, what she calls, classic example of the problems of vaccine surveillance, that of *coincidental events* [62]. In the United States, sudden infant death syndrome (SIDS) during the first year of life occurs at a rate of about 1 in 1,300 infants. One can calculate, she writes, based on age-specific rates of SIDS and the current childhood vaccine schedule, that each year about 50–100 infants can be expected to die of SIDS within 2 days of being vaccinated.

Analysis of surveillance data is associated with statistical problems. Surveillance data contain strong biases. Incidence rates of specific adverse events cannot be calculated. Statistical significance tests and confidence intervals should be used with great reservation. If possible, safety signals should be confirmed in a randomized, controlled clinical trial. Menactra is a meningococcal conjugate vaccine. Meningococcal disease is a potentially fatal infection caused by a bacteria (the meningococcal bacteria), that can infect the blood, the spinal cord and the brain. The vaccine contains four of the most common types of meningococcal bacteria, and was licensed in the United States in 2005 for use in children and adults between the ages of 2 and 55 years old. In September 2005, the FDA and the Center for Biologics Evaluation and Research (CBER) revealed that officials had received five reports of Guillain Barré syndrome (GBS) connected to the Menactra vaccine, all in 17- and 18-year-olds. GBS is a neurological disorder the can cause paralysis and permanent neurological damage. The majority of those affected recover, but recovery may take months and not infrequently may require hospitalization. GBS occurs when the immune system overreacts to foreign invaders. It can occur spontaneously and has been caused by infections, vaccinations, surgical procedures and traumatic injury. GBS was shown to have been a side effect of the swine influenza vaccine during the swine flu outbreak in 1976. To date, post-licensure surveillance did not reveal an association between vaccination with Menactra and GBS.

Post-licensure surveillance is sometimes criticized for underestimating benefits. Almost all children will have had a rotavirus infection by the age of 5. The virus is one of the most common causes of diarrhoea, which can be severe and dehydrating. In developing countries, rotavirus gastroenteritis is a major cause of childhood death. It has been estimated that the infection is responsible for approximately half a million deaths per year among children aged less than 5 years. Rotashield is a live attenuated rotavirus vaccine that was approved by the FDA in 1998. Little more than a year later, the manufacturer voluntarily withdrew it from the market. Shortly after approval, cases of intussusception were reported to the Vaccine Adverse Event Reporting System (VAERS), a surveillance system which collects information about possible side effects of licensed vaccines, a program of the FDA and CDC. Intussusception is a condition in which one bowel segment enfolds within another segment, causing obstruction. After licensure, VAERS recorded 76 cases, with 70% occurring after the first dose of the vaccine. The risk of intussusception has been estimated to be one case in every 5,000–9,500 vaccinated infants. Nonetheless, the vaccine could have prevented a considerable number of deaths in developing countries, where the benefit/risk rate would have been different. But because the vaccine was withdrawn from the United States market it could not be sold in developing countries.

A powerful statistical technique to investigate the association between unwanted events and transient exposure is the *self-controlled case series method*, or case series method for short. The method uses only data on cases, but it can provide estimates of the relative incidence of an adverse event. The method was developed to investigate a possible link between a MMR vaccine used in the United Kingdom and the occurrence of aseptic meningitis (an inflammation of the meninges caused by nonbacterial organisms). Strong evidence for a link between vaccination with the Urabe mumps strain and the disease was found, and several vaccines derived from this genotype mumps strain were withdrawn from the market. The case series analysis is based on conditional maximum likelihood estimation. For every case, a so-called *case series likelihood* is defined, and this likelihood is conditional on the case having occurred during the observation period. The observation period is split into successive intervals determined by changes in covariates and vaccine risk periods. For every period, a Poisson incidence rate is assumed. The case series likelihood is the case's contribution to the Poisson likelihood, conditioned on the case having occurred. The method controls for confounders that do not vary with time such as gender. In 1997, an intranasal influenza vaccine was granted approval for distribution and use in Switzerland. The nasal formulation consisted of an inactivated virosomal influenza vaccine, combined with a powerful mucosal adjuvant, heat-labile *Escherichia coli* enterotoxin. Shortly after the introduction, the vaccine was withdrawn from the market, because it was suspected that use of the vaccine increased the risk of Bell's palsy, a paralysis of the facial nerve leading to an inability to control facial muscles. Often the eye in the affected side cannot be closed and must be protected from drying up, to avoid permanent damage resulting in impaired vision. A report was published in the *New England Journal of Medicine* in which strong evidence for this increased risk was presented [63]. A strong relation in a case-control study was supported by a case series analysis that identified an increase in the incidence of the condition, with a peak occurring between 31 and 60 days after intranasal vaccination followed by a return to the baseline level. For an excellent discussion of the case series method, see the tutorial by Whitaker and colleagues [64].

9.3   Safety Data and the Problem of Multiplicity

The interpretation of safety data is complicated by the problem of multiplicity. The more safety variables are statistically analyzed, the higher the false-positive rate (type I error rate) will be. Also, when the size of the safety database is large, clinically nonrelevant differences will attain statistical significance. But any approach to control the false-positive rate will unavoidably decrease the false-negative rate (type II error rate). This could mean that some adverse vaccine effects may go undetected.

Several approaches to account for multiplicity in the analysis of safety data have been proposed. A first approach is to do nothing, to not correct. For many reviewers of safety data, false negatives (not rejected false safety null hypotheses) are of greater concern than false positives (rejected true safety null hypotheses.) In that case, a conservative approach is not to adjust for multiplicity. This will increase the false-positive rate, but regulatory bodies are aware of this, and such safety signals (flaggings) are rarely a ground for a negative decision. Indeed, whereas regulatory agencies require multiplicity corrections for efficacy data, it is unlikely that they will accept such adjustment for safety data. Safety concerns may get special attention in post-licensure surveillance, or manufactures may be requested to perform a post-marketing safety study.

A second approach is to control the false-positive rate by a multiplicity adjustment that controls the family wise error rate (FWER) in the strong sense, i.e., that controls the probability that at least one true safety null hypothesis is rejected. This can be achieved by applying, for example, the Bonferroni correction method, or the more powerful Holm method. (For a detailed discussion on this correction methods, see the book by Dmitrienko, Tamhane and Bretz [65].) With the Bonferroni correction, if there are m null hypotheses, all hypotheses are tested at the significance level α/m. With the Holm correction, the m P-values are ordered such that

$$P_{(1)} \leq P_{(2)} \leq \cdots \leq P_{(m)}.$$

First $H_{(1)}$ is tested at the level α/m. If $H_{(1)}$ is rejected, then $H_{(2)}$ is tested at the level $\alpha/(m-1)$. If $H_{(2)}$ is rejected, then $H_{(3)}$ is tested at the level $\alpha/(m-2)$, etc. If one of the hypotheses, say, $H_{(i)}$, cannot be rejected then no further null hypotheses are tested. This is why the Holm correction is called a step-wise correction method. But, as already noted, these multiplicity adjustments have the drawback that they increase the false-negative rate. For this reason, this approach is seldom applied in safety data analyses.

A third approach is the *false discovery rate* (FDR) *method*, introduced by Benjamini and Hochberg [66]. The FDR is defined as the expected proportion of rejected safety null hypotheses that are incorrectly rejected. Suppose that a safety analysis involves the statistical testing of 50 independent null hypotheses at the two-sided significance level 0.05, and that 40 of these null hypotheses are true and 10 false. Then the expected proportion of rejected true null hypotheses is $40 \times 0.05 = 2$, while the expected proportion of rejected false null hypotheses is $\sum_i Q_i$, with Q_i the probability that the ith false safety null hypothesis is rejected. If $Q_i = 0.90$ for all 10 false safety null hypotheses, then the expected number of true positives is $10 \times 0.90 = 9$. In that case, the FDR would be $2/11 = 0.18$. (The FDR will be approximately 0.18, because in the calculation the correlation between the numerator and denominator is ignored.) The false discovery rate approach aims to control the FDR at level α, by adjusting the significance level at which the safety null hypotheses are tested. When all safety null hypothesis are true, then the FDR procedure controls the family wise error rate in the strong sense. But when some safety null hypotheses are false, then the statistical power of the FDR approach exceeds that of methods that control the FWER. Let

$$P_{(1)} \leq P_{(2)} \leq \cdots \leq P_{(m)}$$

be the ordered P-values for testing the safety null hypotheses

$$H_{(1)}, H_{(2)}, \ldots, H_{(m)}.$$

The FDR procedure rejects the j null hypotheses $H_{(1)}, H_{(2)}, \ldots, H_{(j)}$, where

$$j = \max\{i : P_{(i)} \leq (i/m)\alpha\}.$$

Table 9.1 Comparison of three approaches to control for multiplicity

Correction method	$P_{(1)}$	$P_{(2)}$	$P_{(3)}$	$P_{(4)}$	$P_{(5)}$
	0.0045*	0.0120	0.0212	0.0224	0.0493
Uncorrected	+	+	+	+	+
Bonferroni	+	−	−	−	−
Holm	+	+	−	−	−
FDR	+	+	+	+	+
	0.0225**	0.0280	0.0280	0.0280	0.0493

* Unadjusted P-value; ** Adjusted P-value; *Plus*: Corresponding null hypothesis rejected; *Minus*: Null hypothesis not rejected

The adjusted P-values for the FDR procedure are:

$$\text{adjusted } P_{(m)} = P_{(m)}$$

$$\text{adjusted } P_{(j)} = \min\{\text{adjusted } P_{(j+1)}, (m/j)P_{(j)}\}, \quad \text{for } j < m.$$

Consider the five ordered P-values in Table 9.1. All uncorrected P-values are < 0.05, and thus all five safety null hypotheses $H_{(1)}$, $H_{(2)}, \ldots, H_{(5)}$ would be rejected. When the Bonferroni method is applied, all null hypotheses have to be tested at the $0.05/5 = 0.01$ significance level, and in that case only $H_{(1)}$ would be rejected. When the Holm method is applied, $H_{(1)}$ must be tested at the level $0.05/5 = 0.01$, $H_{(2)}$ at the level $0.05/4 = 0.0125$, and $H_{(3)}$ at the level $0.05/3 = 0.0167$. Because $P_{(3)} = 0.0212 > 0.0167$, $H_{(3)}$ cannot be rejected, and because the Holm method is a step-wise procedure, $H_{(4)}$ and $H_{(5)}$ can also not be rejected. With the Bonferroni method only one null hypothesis would be rejected, while with the Holm method two null hypotheses would be rejected, which illustrates the difference in power between the two methods. On the last row of Table 9.1, the adjusted P-values for the FDR method are shown. All adjusted P-values are <0.05 and thus all five null hypotheses would be rejected. To get a feeling for the differences between the uncorrected approach, the Holm method and the FDR approach, consider a study in which 45 independent safety null hypotheses are tested at the 0.05 significance level, and that 40 of these null hypotheses are true and 5 false, and that for each of these 5 null hypotheses the probability of a true-positive result is 0.90. Ten-thousand studies were simulated, and the results are shown in Table 9.2. If no corrections are made, the probability of at least one false positive (the probability of rejection at least one true safety null hypothesis) is as high as 0.873. The expected number of false positives is 2.0, while the expected number of true positives is 4.5. If the Holm method is applied, the probability that at least one true null hypothesis is rejected is 0.047. The expected number of false positives is 0.048, while the expected number of true positives is 2.3. Finally, if the FDR method is applied, the probability of at least one false positive is 0.175. The expected number of false positives is 0.02, and the expected number of true positives is 3.0. The FDR is 0.045, a value which is indeed smaller than the significance level 0.05. The FDR method is thus a

Table 9.2 Results of 10,000 simulated safety studies

Multiplicity correction method	FWER[a]	Average number false positives	Average number true positives	FDR
No correction	0.875	2.0	4.5	0.281
Holm	0.047	0.05	2.3	0.017
FDR	0.169	0.20	3.0	0.045

[a] Average number of studies with at least one false positive
(= At least one rejected true safety null hypothesis)

compromise between the uncorrected method – less false positives – and the Holm method – more true positives.

A fourth approach was proposed Mehrotra and Heyse, the *double false discovery rate* approach [67]. The double FDR approach is a two-step procedure for flagging adverse events. First, adverse events are grouped by body systems (e.g., the Body Systems of MedDRA, the standard medical terminology designed for the classification of medical information throughout the medical product regulatory cycle.) Suppose there are s body systems, and let P_{ik} be the P-value for testing H_{ik}, the kth safety null hypothesis of the ith body system. Then

$$P_i* = \min\{P_{i1}, P_{i2}, \ldots, P_{ik}\}$$

is the 'representative' P-value for the ith body system, i.e., the P-value for the strongest safety signal. The FDR procedure is applied to the P_i* and, within the body systems, to the $P_{i1}, P_{i2}, \ldots, P_{ik}$. The double FDR procedure flags H_{ik} if

$$\text{adjusted } P_i* < \alpha_1 \text{ and adjusted } P_{ik} \leq \alpha_2.$$

The authors advice to set α_1 to $\alpha/2$ and α_2 to α if the FDR is to controlled at level α. The Double FDR method substantially reduces the percentage of incorrectly flagged adverse events because it takes (some of) the dependency between events into account.

A fifth procedure that needs to be mentioned here was proposed by Berry and Berry [68]. Their approach is a Bayesian alternative to the double FDR approach, and since its publication in 2004 it has gained considerable popularity. Their model is a three-level hierarchical mixture model for simultaneously addressing many types of adverse events that are, like in the double FDR approach, grouped into body systems. The strength of the model is that it allows borrowing information both across and within body systems. Because of its complexity, the model is not discussed here. However, the publication may be of special interest to statisticians working in vaccine research because it presents a re-analysis of the vaccine safety data of Mehrotra and Heyse.

9.4 Vaccine Reactogenicity

With vaccine *reactogenicity* is meant common adverse events that are considered to be caused by or be attributable to the vaccination. They can be local or systemic. The reactions to be assessed depend on the type (class) of vaccine, its mechanism of action, route of administration, the targeted disease and the target population. Pre-licensing safety analyses typically focus on reactogenicity data.

9.4.1 Local and Systemic Reactions

Local reactions are reactions that occur at the site where the vaccine is administrated. In case of an injectable vaccine, these reactions are often called *injection site reactions*. They can be caused either by needle trauma or as an inflammatory reaction to the vaccine constituents. For example, local injection site pain is usually the consequence of some degree of tissue damage. Other examples of injection site reactions are: impairment of arm movement, tenderness, erythema (redness), induration (swelling), itching, ecchymosis (blue spots). Common local reactions after nasal vaccination are nasal congestion and runny nose. A special class of local reactions are adjuvant-related local reactions, and the reader is referred to Chap. 18 of the book edited by M. Singh for a discussion on this topic [69].

Local reactions may be accompanied by *systemic reactions*, which are reactions that are the result of the immunological response to the vaccine. Typical examples of systemic reactions are: headache, fever, malaise, fatigue, arthralgia (noninflammatory joint pain), myalgia (muscle pain) and increased sweating. When the target population are toddlers, often graded systemic reactions are: crying, irritability and decreased feeding.

Local and systemic reactions are usually collected with the help of a diary, which has to be filled in by the subject, or the parents in case of a childhood vaccine, for a period of 3 or 7 days after the vaccination. If reactions are predefined on the diary, they are called *solicited reactions* (in contrast to the *unsolicited other adverse events*, which are collected on the adverse events pages of the case report form).

Severity of local and systemic reactions can be graded on a binary scale (yes, no), but more often an ordinal (ordered categorical) scale with the following 4 categories is used: *none, mild, moderate, severe*. A standard *functional grading* (categorization) is

    mild: not interfering with normal daily activities
moderate: interfering with normal daily activities
  severe: preventing one or more normal daily activities

This grading makes the scale suited for rating reactions by subjects, which is an attractive property, but it may not be very sensitive to differences between vaccines. Also, it is not suited to grading a systemic reactions such as fever. In an attempt to introduce uniform criteria for grading reactions, FDA/CBER in 2007 published grading scales for clinical and laboratory abnormalities for preventive vaccine clinical trials [70]. For erythema the grades are based on the size of the greatest single diameter, while for induration (hardening of the skin) the grades are based on a functional assessments as well as an actual measurement. The FDA/CBER grades for the systemic reactions such as nausea and vomiting also take into account the number of episodes, while the grades for headache account for the use of nonnarcotic and narcotic pain relievers. The grades for diarrhoea are based on the frequency, shape and weight of the stools. (The FDA/CBER grading scales all contain a fourth grade: potentially life threatening. Because this grade will rarely occur, it is usually omitted.)

9.4.2 Statistical Analysis of Local and Systemic Reactions

The standard statistical analysis of reactogenicity data assesses the incidence, the severity, the duration and, sometimes, the time after the vaccination of the local and systemic reactions.

9.4.2.1 Analyzing Incidences of Local and Systemic Reactions

The statistical analysis of local and systemic reactions usually starts with quantifying the incidences of the individual reactions by means of confidence intervals. Here, with incidence is meant the proportion of subjects reporting the reactions at least once during the 3 or 7 days follow-up period. The focus will usually be on the upper confidence limit for the incidence, because it gives an upper bound for the rate with which the reaction is expected to occur among subjects receiving the vaccine. The bound is often translated into a *less-than-1-in rate*. If the upper confidence limit for the incidence of a specific reaction is CL_U, then the expected rate of the reaction is <1 in $1/CL_U$ vaccinated subjects, with $1/CL_U$ often rounded down to the nearest multiplier of 100.

Example 9.1. Consider a vaccine safety database of 4,500 subjects who received the vaccine. Suppose that the systemic reaction sinusitis (inflammation of the paranasal sinuses) was reported for 3 subjects. The upper limit of the 95% Clopper–Pearson confidence interval for the incidence of sinusitis is 0.0019. Thus, the expected rate of sinusitis is <1 in 526 (i.e., <1 in 500) vaccinated subjects.

When reactogenicity experiences are to be compared between two vaccines – e.g., between an investigational vaccine and a control vaccine – it is done by computing confidence intervals for the relative risks.

Example 9.2. In Table 9.3, the observed incidences of 4 selected systemic reactions reported by Mehrotra and Heyse (see Sect. 9.3) are given. Shown are the observed incidences of the reactions malaise, constipation, diarrhoea and urticaria (hives)

Table 9.3 Incidences of 4 systemic reactions in a measles, mumps, rubella and varicella vaccines trial

Systemic reaction	MMRV group ($n = 148$)	MMR + V group ($n = 132$)	Relative risk ρ^a
Malaise	27	20	1.16^b (0.72, 2.03)
Constipation	2	0	1.80
Diarrhoea	24	10	1.96 (1.09, 4.27)
Urticaria	0	2	0.00

[a] MMRV versus MMR + V; [b] with Jewell's correction

for the MMRV and the MMR + V vaccination groups. Also shown are the points estimates and the two-sided 95% Wilson-type confidence intervals for the risk ratios.

Note that when one of the number of cases is zero the Wilson-type confidence interval for the relative risk cannot be evaluated.

Comparing incidences of local or systemic reactions between two vaccine groups is straightforward, but the disadvantage of comparing individual reactions between vaccine groups is that it does not allow accumulation of evidence. Such evidence could, for example, be that for all local or for all systemic reactions the rate ratio exceeded 1.0, but with none of the P-values being significant. Thus, although the data would strongly suggest that vaccine A is more reactogenic than vaccine B, there would be no statistical evidence to claim this. In that case, a simple but powerful approach is to analyze the intra-individual total numbers of local or systemic reactions. If on the diary there are, say, six solicited local reactions, then the intra-individual total number of local reactions can be 0, 1, 2, 3, 4, 5 or 6. Total numbers can be compared between two vaccine groups by means of Wilcoxon's rank-sum test.

Example 9.3. De Bruijn and co-workers compared the reactogenicity of a virosomal influenza vaccine to that of an MF59-adjuvanted vaccine in elderly [71]. The number of solicited local reactions on the diary was eight. Below the SAS analysis of the total numbers of local reactions is given. Note that the analysis was done using procedure FREQ (rather than procedure NPAR1WAY), using the cmh option with rank scores.

SAS Code 9.1 *Comparing Intra-Individual Numbers of Local Reactions*

```
proc freq;
    table vaccine*number_of_local_reactions /
        nopercent nocol chm scores=rank;
run;
```

SAS Output 9.1

```
VACCINE      NUMBER_OF_LOCAL_REACTIONS

Frequency
Row Pct         0      1      2      3      4      5      6      7 Total
---------- ------ ------ ------ ------ ------ ------ ------ ------

ADJUVANTED     70     26     13      8      3      5      1      4   130
            53.85  20.00  10.00   6.15   2.31   3.85   0.77   3.08
---------- ------ ------ ------ ------ ------ ------ ------ ------

VIROSOMAL     100     18      7      1      0      0      0      1   127
            78.74  14.17   5.51   0.79   0.00   0.00   0.00   0.79
---------- ------ ------ ------ ------ ------ ------ ------ ------

Total         170     44     20      9      3      5      1      5   257


     Cochran-Mantel-Haenszel Statistics (Based on Rank Scores)

Statistic    Alternative Hypothesis     DF      Value       Prob
-----------------------------------------------------------------
     1       Nonzero Correlation         1     21.1839    <.0001
     2       Row Mean Scores Differ      1     21.1839    <.0001
     3       General Association         7     24.6651    0.0009
```

On average, the subjects vaccinated with the adjuvanted vaccine reported more local reactions than the subjects vaccinated with the virosomal vaccine. In the adjuvanted vaccine group, 43.2% of the subjects reported at least one local reaction, and 16.2% reported three or more reactions. In contrast, in the virosomal group only 21.2% of the subjects reported at least one local reaction while only 2 subjects reported 3 or more reactions. The statistic to compare the intra-individual total number is local reactions is Statistic 2, which is a Chi-square statistic with 1 degree of freedom. For the example data the two-sided P-value is < 0.001, which allowed the conclusion that in elderly the adjuvanted influenza vaccine is more reactogenic than the virosomal vaccine with respect to local reactions.

9.4.2.2 Analyzing the Severity of Local and Systemic Reactions

As said, local and systemic reactions are usually scored for a period of 3 or 7 days after the vaccination. In that case, the intensity of the reaction is usually taken to be the maximum score during the follow-up period.

Local and systemic reaction ordinal scores can be compared between two vaccine groups by means of Wilcoxon's rank-sum test with modified ridit scores as ranks [72]. The category *none* is given grade 0, the category *mild* grade 1, the category *moderate* grade 2 and the category *severe* grade 3. The null hypothesis tested is that mean scores do not differ between the groups. This comparison can be done with procedure FREQ of SAS, using the chm option with modridit scores.

Example 9.4. According to the results of a randomized study published in the *British Medical Journal*, longer needles for infant immunizations may cause fewer local reactions [73]. Compared with short narrow needles, use of long wide needles was associated with significantly decreased local reactions to diphtheria, tetanus, whole cell pertussis, and *H. influenzae* type b vaccinations. Significantly fewer infants vaccinated with the long needle had severe local reactions. Consider a (hypothetical) randomized trial comparing administration of a diphtheria vaccine using either a long (25 mm) needle or short (16 mm) needle. Suppose that local reactions were graded by parents trained how to do so, and that for the local reaction tenderness the results were as follows: infants vaccinated with the long needle: *none*: 30, *mild*: 20, *moderate*: 12, *severe* tenderness: 5; infants vaccinated with the short needle: *none*: 19, *mild*: 15, *moderate*: 19, *severe*: 10. To analyze these reaction scores, the following SAS code can be used:

SAS Code 9.2 *Comparing Ordinal Reaction Scores*

```
proc freq;
   table vaccine*tenderness_score / nopercent nocol chm scores=modridit;
run;
```

SAS Output 9.2

```
The FREQ procedure

VACCINE      TENDERNESS_SCORE
Frequency
Row Pct          0        1        2        3    Total
---------  -------- -------- -------- --------
A               30       20       12        5       67
             44.78    29.85    17.91     7.46
---------  -------- -------- -------- --------
B               19       15       19       10       63
             30.16    23.81    30.16    15.87
---------  -------- -------- -------- --------
Total           49       35       31       15      130

Summary Statistics for VACCINE by TENDERNESS_SCORE

   Cochran-Mantel-Haenszel Statistics (Modified Ridit Gradings)

 Statistic   Alternative Hypothesis     DF       Value       Prob
 ------------------------------------------------------------------
     1        Nonzero Correlation        1       5.5634      0.0183
     2        Row Mean Scores differ     1       5.5634      0.0183
     3        General Association        3       6.2653      0.0994
```

Here also, the statistic to use is Statistic 2. For the data in Example 9.6 the two-sided P-value is 0.0183. In the database the scores must be ordered, e.g., 0, 1, 2, 3 or A, B, C, D but not 'none', 'mild', 'moderate', 'severe', because in that case procedure FREQ ranks the scores alphabetically, i.e., as: 'mild' = 1, 'moderate' = 2, 'none' = 3, 'severe' = 4, in which case an incorrect P-value is returned: 0.7272.

Appendix A
SAS and Floating Point Format
for Calculated Variables

When using SAS, floating point format for calculated variables should be avoided, especially when values are to be compared with a constant. As shown below, it may lead to errors. The solution to this problem is rounding. When calculating a value use the function ROUND at the final step and round to, say, three decimals more than needed for the comparison. (But do not round to soon.) As an example, consider a trial in which every serum sample is titrated twice, with the titre assigned to the sample the geometric mean of the two assay values. Let the endpoint be whether or not the subject is seroprotected, e.g., whether or not the assigned titre is greater or equal to 40. With the floating point format errors will occur.

SAS Code A.1

```
data;
   input subject assay1 assay2;
   titre=exp((log(assay1)+log(assay2))/2);      /* geometric mean */
   titre_r=round(titre,.001);      /* rounded titre              */
   sp=(titre ge 40);               /* seroprotected yes/no       */
   sp_r=(titre_r ge 40);           /* seroprotected yes/no derived */
datalines;                         /* from rounded titre         */
1 40  40
2 20  80
3 10 160
4  5 320
run;

proc print; run;
```

SAS Output A.1

subject	assay1	assay2	titre	titre_r	sp	sp_r
1	40	40	40	40	1	1
2	20	80	40	40	0	1
3	10	160	40	40	1	1
4	5	320	40	40	1	1

All assigned titres should be 40, and for all four subjects both the nonrounded and the rounded calculated titre is printed as 40. But, when the calculated titre is not rounded, according to SAS, subject 2 is not seroprotected. This is due to the use of the floating point format. When the values are rounded this error does not occur.

Appendix B
Closed-Form Solutions for the Constrained ML Estimators $\tilde{R}_0$

The standard errors (3.8) and (3.11) involve constrained maximum likelihood estimators $\tilde{R}_0$ and $\tilde{R}_1$ of the rates π_0 and π_1. For standard error (3.8), the constraint is

$$\tilde{R}_1 - \tilde{R}_0 = \Delta.$$

Let s_0 and s_1 be the observed numbers of events, n_0 and n_1 the group sizes, $s = s_0 + s_1$ and $n = n_0 + n_1$. Define:

$$L_0 = s_0 \Delta (1 - \Delta)$$
$$L_1 = (n_0 \Delta - n - 2s_0)\Delta + s$$
$$L_2 = (n_1 + 2n_0)\Delta - n - s.$$

The closed-form solution for $\tilde{R}_0$ is

$$\tilde{R}_0 = 2p \cos(a) - L_2/(3n),$$

where

$$a = (1/3)[\pi + \cos^{-1}(q/p^3)]$$
$$q = L_2^3/(3n)^3 - L_1 L_2/(6n^2) + L_0/(2n)$$
$$p = \text{sign}(q)\sqrt{L_2^2/(3L_3)^2 - L_1/(3n)}.$$

For standard error (3.11) the constraint is

$$\tilde{R}_1 = \theta \tilde{R}_0.$$

The closed-form solution for $\tilde{R}_0$ is

$$\tilde{R}_0 = \frac{-B - \sqrt{B^2 - 4AC}}{2A},$$

where

$$A = n\theta$$
$$B = -(n_1 \theta + s_1 + n_0 + s_0 \theta)$$
$$C = s.$$

Appendix C
Simulation Results on Jewell's Correction for the Rate Ratio

Jewell's correction is a simple but powerful correction to remove the bias in the standard relative risk estimator, the rate ratio:

$$RR = \frac{s_1/n_1}{s_0/n_0},$$

with s_1 and s_0 the observed numbers of events and n_1 and n_0 the group sizes. The bias will be nonnegligible when the control rate π_0 is close to zero and n_0 is small to intermediate. Jewell's correction is to set s_0 to $(s_0 + 1)$ and n_0 to $(n_0 + 1)$.

In Table C.1, simulation results on the performance of Jewell's correction are shown. For selected combinations (n_0, π_0, n_1, π_1), 5,000 pairs of random samples were drawn, one from the binomial distribution $BIN(n_0, \pi_0)$ and one from the binomial distribution $BIN(n_1, \pi_1)$. For each pair both the uncorrected and the corrected rate ratio was calculated. The rate ratios shown in the table are the averages of the 5,000 simulated ratios. The simulations confirm the bias of the standard estimator, it overestimates $\theta = \pi_1/\pi_0$. The performance of the corrected estimator is excellent.

Table C.1 Monte Carlo simulation results on the performance of the standard and Jewell's corrected rate ratio

π_0	π_1	n_0	n_1	θ	RR	$RR_{corr.}$
0.1	0.05	50	50	0.5	0.68	0.50
0.1	0.1	50	50	1.0	1.26	0.99
0.1	0.2	50	50	2.0	2.53	1.99
0.1	0.05	50	200	0.5	0.63	0.50
0.1	0.1	50	200	1.0	1.27	1.00
0.1	0.2	50	200	2.0	2.48	1.95
0.1	0.05	200	50	0.5	0.57	0.50
0.1	0.1	200	50	1.0	1.05	0.99
0.1	0.2	200	50	2.0	2.10	2.00
0.1	0.05	200	200	0.5	0.53	0.50
0.1	0.1	200	200	1.0	1.05	0.99
0.1	0.2	200	200	2.0	2.09	2.00
0.3	0.3	50	50	1.0	1.06	1.01

Appendix D
Proof of Inequality (3.16)

Consider a trial with $k > 1$ co-primary endpoints, and with the objective to demonstrate that an experimental vaccine is superior (or noninferior) to a control vaccine for *all* co-primary endpoints. Let E_i be the event that the trial yields a significant result for the ith endpoint, $P_i = \Pr(E_i)$ the statistical power of the trial for the ith endpoint, and $P = \Pr(E_1 \cap \ldots \cap E_k)$ the overall statistical power, i.e., the probability that the trial yields a significant result for all k endpoints. Then the following inequality holds

$$P \geq \sum_{i=1}^{k} P_i - (k - 1).$$

Proof. The inequality can be proven by mathematical induction. According to the addition rule for probabilities:

$$
\begin{aligned}
P &= \Pr(E_1 \cap E_2) \\
&= \Pr(E_1) + \Pr(E_2) - \Pr(E_1 \cup E_2) \\
&\geq \Pr(E_1) + \Pr(E_2) - 1 \\
&= P_1 + P_2 - (2 - 1).
\end{aligned}
$$

Thus, the inequality holds for $k = 2$. Assume that it has been shown that the inequality holds for $k = 2, \ldots, j$, with $j \geq 2$. Then for $k = (j + 1)$ it follows that

$$
\begin{aligned}
P &= \Pr(E_1 \cap \cdots \cap E_j \cap E_{j+1}) \\
&= \Pr(E_1 \cap \cdots \cap E_j) + \Pr(E_{j+1}) - \Pr((E_1 \cap \cdots \cap E_j) \cup E_{j+1}) \\
&\geq \Pr(E_1 \cap \cdots \cap E_j) + \Pr(E_{j+1}) - 1 \\
&\geq P_1 + \cdots + P_j - (j - 1) + \Pr(E_{j+1}) - 1 \\
&= P_1 + \cdots + P_{j+1} - j \\
&= P_1 + \cdots + P_{j+1} - [(j + 1) - 1].
\end{aligned}
$$

$\square$

Appendix E
A Generalized Worst-Case Sensitivity Analysis for a Single Seroresponse Rate for Which the Confidence Interval Must Fall Above a Pre-Specified Bound

E.1 Introduction

In 2007, FDA/CBER published two guidance documents for the licensure of influenza vaccines, one for seasonal inactivated vaccines and another for pandemic vaccines [30, 31]. Both documents give the same criteria for influenza vaccine immunogenicity. For an adult population, the lower limit of the two-sided 95% confidence interval for the seroprotection rate must meet or exceed 0.7, and the lower limit of the confidence interval for the seroconversion rate must meet or exceed 0.4. For an elderly population, the respective bounds are 0.6 and 0.3. Seroprotection and seroconversion are both binary outcomes. Seroprotection is defined as achieving an antibody level above a given threshold value. The standard definition of seroconversion is going from a pre-vaccination state of no detectable antibodies (seronegative) to a post-vaccination state of detectable antibodies (seropositive). An alternative definition of seroconversion is a significant post-vaccination increase in antibody level.

In case of a statistical analysis aimed at demonstrating that the confidence interval of a rate is above a pre-specified bound, the most applied method to handle missing data is, probably, the complete-case analysis. This analysis requires the assumption that the probability that an outcome is missing is independent of the outcome, i.e., the assumption that the probability that the outcome is missing does not depend on whether the outcome is positive (success, e.g., subject seroconverted) or negative (failure). A sensitivity analysis is an analysis that investigates the influence of deviations from the assumptions underlying the main analysis. For binary outcomes, a simple sensitivity analysis in case of missing data is to treat all subjects with a missing outcome as failures, and then to check if this analysis supports the conclusion from the complete-case analysis. This analysis, that is, the worst-case sensitivity analysis – is based on an extreme assumption, namely that only failures will be missing, and the more missing data there are, the more extreme the assumption is.

Here, a generalized worst-case sensitivity analysis for a single rate for which the confidence interval must fall above a pre-specified bound is proposed, based on the maximum likelihood (ML) method. This generalized analysis checks for a continuum of assumptions, from the assumption underlying the complete-case analysis to the one underlying the worst-case analysis, if the bound lies within or outside the confidence interval.

E.2 Motivating Example

As a motivating example, consider a study in which 100 adult subjects are vaccinated with a pandemic A-H1N1 influenza vaccine. Suppose that three weeks after the vaccination 49 subjects have seroconverted and 41 not, and that for 10 subjects the outcome is missing. In the complete-case analysis, the FDA/CBER criteria for seroconversion is met because the lower limit of the 95% Clopper–Pearson confidence interval for the probability of seroconversion is 0.436, which exceeds the bound set by the agency, $\delta = 0.4$. A sensitivity analysis in which all subjects with a missing outcome are assumed to have not seroconverted, however, does not support the conclusion of the complete-case analysis because in that analysis the lower limit of the Clopper–Pearson confidence interval is 0.389.

E.3 Complete-Case and Worst-Case Maximum Likelihood Analyses

The generalized worst-case sensitivity analysis proposed here is based on the ML method. Therefore, as an introduction, first the ML analyses of the complete-case and the worst-case data are described.

Let θ denote the probability of a positive outcome, and π_s and π_{ns} the probabilities that a positive or a negative outcome is missing. In Table E.1, a probability model for the data including missing values is given. Note that it is assumed that the missing data mechanism depends on the outcome but not on any observed or nonobserved covariate. The log-likelihood function for the data set is

$$LL(\theta, \pi_s, \pi_{ns}) = s \log[(1 - \pi_s)\theta] + (m - s) \log[(1 - \pi_{ns})(1 - \theta)] \quad \text{(E.1)}$$
$$+(n - m) \log[\pi_s \theta + \pi_{ns}(1 - \theta)],$$

with s the observed number of positive outcomes, m the total number of subjects with a nonmissing outcome and n the total number of subjects.

Table E.1 Probability model

Event	Probability
Observed positive outcome	$(1 - \pi_s)\theta$
Observed negative outcome	$(1 - \pi_{ns})(1 - \theta)$
Missing observation	$\pi_s \theta + \pi_{ns}(1 - \theta)$

The complete-case analysis requires the assumption that $\pi_s = \pi_{ns} = \pi$. In that case the function in (E.1) becomes

$$LL(\theta, \pi) = [s \log \theta + (m - s) \log(1 - \theta)] + [(n - m) \log \pi + m \log(1 - \pi)].$$

The first component of this log-likelihood function depends only on θ while the second component depends only on π. Thus, both components can be maximized independently of each other. If the parameter of interest is θ, then the second component is a constant and can be dropped from the log-likelihood function, which then simplifies to the log-likelihood function for complete-case data:

$$LL_{CC}(\theta) = s \log \theta + (m - s) \log(1 - \theta).$$

The null hypothesis $H_0: \theta = \theta_0$ can be tested using the likelihood ratio statistic:

$$LRS_{CC}(\theta_0) = 2[LL_{CC}(\hat{\theta}) - LL_{CC}(\theta_0)],$$

where $\hat{\theta}$ is the ML estimate of θ. Under the null hypothesis, for large sample sizes this statistic has a Chi-square distribution with one degree of freedom. The likelihood ratio statistic can be used to derive a confidence interval for θ. Any value θ_0 for which $LRS_{CC}(\theta_0)$ is less than $\chi^2_{1-\alpha}$ is in the $100(1 - \alpha)\%$ likelihood-based confidence interval, and *vice versa*.

Example. (continued) For the complete-case data, $\hat{\theta}$ is $49/90 = 0.544$, with LL_{CC} $(\hat{\theta}) = -62.027$. For the FDA/CBER bound for seroconversion the log-likelihood equals $LL_{CC}(0.4) = -65.842$. Thus, $LRS_{CC}(0.4) = 7.630 > \chi^2_{0.95} = 3.841$, which implies that the bound is not in the 95% confidence interval. The lower confidence limit has to be found by iteration. $LRS_{CC}(0.442) = 3.796 < \chi^2_{0.95}$ and $LRS(0.441) = 3.971 > \chi^2_{0.95}$. For the complete-case data, the lower likelihood-based confidence limit for the probability of seroconversion is 0.442, which is in good agreement with the Clopper–Pearson limit.

The log-likelihood function for the worst-case data, i.e., for the data set with the missing values replaced by zeros (failures), is

$$LL_{WC}(\theta) = s \log \theta + (n - s) \log(1 - \theta). \tag{E.2}$$

This is the same log-likelihood function as for the complete-case data, except that in the second term the multiplier $(m - s)$ is now $(n - s)$.

Example. (continued) For the worst-case data, $\hat{\theta}$ is $49/100 = 0.490$, with $LL_{WC}(\hat{\theta}) = -69.295$, $LL_{WC}(0.4) = -70.950$, and $LRS_{WC}(0.4) = 3.311 < 3.841$. Again, the likelihood analysis is in agreement with the Clopper–Pearson analysis, that for the worst-case data the FDA/CBER bound is not below but in the 95% confidence interval.

E.4   Maximum Likelihood Analysis with Missing Data

With the following re-parameterization: $\lambda = \pi_{ns}/\pi_s$ and $\pi = \pi_{ns}$, the log-likelihood function in (E.1) becomes

$$LL(\theta, \pi, \lambda) = s \log[(1 - \pi/\lambda)\theta] + (m - s) \log[(1 - \pi)(1 - \theta)]$$
$$+(n - m) \log[(\pi/\lambda)\theta + \pi(1 - \theta)].$$

Let $\tilde{\theta}_\lambda$ and $\tilde{\pi}_\lambda$ denote the constrained ML estimates of θ and π for λ fixed. The conditional null hypothesis H_0: $(\theta = \theta_0|\lambda)$ can be tested using the conditional likelihood ratio statistic:

$$\text{CLRS}(\theta_0|\lambda) = 2[LL(\tilde{\theta}_\lambda, \tilde{\pi}_\lambda, \lambda) - LL(\theta_0, \tilde{\pi}_{0\lambda}, \lambda)], \qquad (E.3)$$

where $\tilde{\pi}_{0\lambda}$ is the constrained ML estimate of π under the conditional null hypothesis. The statistic $\text{CLRS}(\theta_0|\lambda)$ can be considerably simplified, because the log-likelihood LL has an interesting property, namely that $\tilde{\pi}_{0\lambda} = \tilde{\pi}_\lambda$. A proof of these property is given below (see *Technical Notes*). Because of this property an alternative formula for the statistic is

$$\text{CLRS}(\theta_0|\lambda) = 2[LL'(\tilde{\theta}_\lambda, \lambda) - LL'(\theta_0, \lambda)], \qquad (E.4)$$

with (see formula (E.6))

$$LL'(\theta, \lambda) = [s \log \theta + (m - s) \log(1 - \theta) + (n - m) \log(\theta/\lambda + 1 - \theta)]. \quad (E.5)$$

The formula in (E.4) is much easier to evaluate than that in (E.3) because it does not involve $\tilde{\pi}_\lambda$. Furthermore, for $\tilde{\theta}_\lambda$ a closed-form solution exists (see *Technical Notes*). Thus, to evaluate $\text{CLRS}(\theta_0|\lambda)$, no (iterative) maximization is required.

Under the conditional null hypothesis, $\text{CLRS}(\theta_0|\lambda)$ has a Chi-square distribution with one degree of freedom. It is easy to see that when λ is set to 1.0, the complete-case analysis is obtained, and that in that case $\tilde{\theta}_\lambda = \hat{\theta}$ (i.e., the ML estimate for the complete-case analysis) and $\text{CLRS}(\theta_0|1.0) = LRS_{CC}(\theta_0)$.

E.5   Generalized Sensitivity Analysis

A generalized worst-case sensitivity analysis is to inspect for which values for the sensitivity parameter λ the lower limit of the constrained likelihood-based confidence interval meets or exceeds δ. This is done by testing the conditional null hypothesis H_0: $(\theta \leq \delta|\lambda)$ for successive values for λ at the one-sided 0.025 significance level.

Table E.2 Generalized worst-case sensitivity analysis of the example data

λ	$\tilde{\tilde{\theta}}_\lambda$	$LL'(\tilde{\tilde{\theta}}_\lambda, \lambda)$	$LL'(0.4, \lambda)$	CLRS$(0.4|\lambda)$
1.0	0.544	−62.027	−65.842	7.630
2.0	0.526	−65.140	−68.074	5.868
3.0	0.516	−66.389	−68.944	5.110
4.0	0.511	−67.062	−69.409	4.694
5.0	0.507	−67.482	−69.699	4.434
6.0	0.505	−67.769	−69.897	4.256
7.0	0.503	−67.978	−70.041	4.126
8.0	0.501	−68.137	−70.150	4.026
9.0	0.500	−68.261	−70.236	3.950
10.0	0.499	−68.361	−70.305	3.888
11.0	0.498	−68.444	−70.362	3.836
12.0	0.498	−68.513	−70.410	3.794
∞	0.490	−69.295	−70.950	3.310

Example. (continued) In Table E.2, results are shown for selected values for λ. The null hypothesis H_0: $\theta \leq 0.4$ is rejected for values for λ as large as 10.0. Thus, even under the extreme assumption that the probability that the outcome of a nonseroconverted is missing is ten times as high as the probability that the outcome of a seroconverted subject is missing, the data supports the conclusion that $\theta > 0.4$. Only if a more extreme value for λ is assumed, the conclusion from the complete-case analysis is not supported. This can be compared with the reasons why the outcomes are missing.

In the worst-case analysis π_s is assumed to be to 0.0, meaning that λ is assumed to be ∞. In that case, the log-likelihood function LL' in (E.5) simplifies to

$$LL'(\theta, \lambda) = s \log \theta + (n - s) \log(1 - \theta).$$

This is the log-likelihood function for the worst-case data (see (E.2)). Thus, the log-likelihood analysis of the worst-case data yields identical results as the log-likelihood analysis with missing values λ set to ∞. This shows that the worst-case analysis is the limiting case of the generalized worst-case sensitivity analysis. The generalized analysis thus has the following nice property: if the worst-case analysis supports the complete-case analysis, so will the generalized analysis; if the worst-case analysis does not support the complete-case analysis, neither will the generalized analysis for larger values for λ.

E.6 Concluding Remarks

The advantage of the generalized worst-case sensitivity analysis is that the robustness of the complete-case analysis can be checked for less extreme assumptions than the assumption that the sensitivity parameter λ is infinite. This is a considerable gain because the assumption that λ can be infinite will rarely be realistic. Consider again the example. Suppose that four values were missing because the tube with the serum sample was broken during transport, two due to loss-to-follow-up and four because the analysis of the serum sample failed. The first reason can be assumed to be unrelated to the antibody level, but suppose that it is known that a failed serum sample analysis is more likely to occur for low antibody levels. If it is further assumed that loss-to-follow-up may also correlate with a low antibody level, then the expected number of missing positive outcomes is 2, and the expected number of missing negative outcomes 8. In that case an estimate of the probability of a missing positive outcome is 2/43, and an estimate of the probability of a missing negative outcome is 8/57. Thus, an estimate of λ would be $(8/57)/(2/43) = 3.0$. For this and comparable values for λ, the generalized worst-case sensitivity analysis supported the conclusion of the complete-case analysis.

E.7 Technical Notes

The log-likelihood function in (E.2) can be factorized as

$$
\begin{aligned}
LL(\theta, \pi, \lambda) &= [s \log \theta + (m - s) \log(1 - \theta) + (n - m) \log(\theta/\lambda + 1 - \theta)] \\
&+ [s \log(1 - \pi/\lambda) + (m - s) \log(1 - \pi) + (n - m) \log \pi]. \quad \text{(E.6)}
\end{aligned}
$$

With λ fixed, both components can be maximized independently, meaning that constraint ML estimates of π are independent of θ. This implies that $\tilde{\pi}_{0\lambda} = \tilde{\pi}_\lambda$.

Differentiating the log-likelihood function in (E.5) yields the following normal equation

$$
\frac{s}{\theta} - \frac{(m - s)}{1 - \theta} + \frac{(n - m)\lambda'}{1 + \theta\lambda'} = 0,
$$

with $\lambda' = (1/\lambda - 1)$. Solving this equation for θ produces the constrained ML estimate $\tilde{\theta}_\lambda$. Simple algebra yields that for $\lambda > 1.0$ the estimate $\tilde{\theta}_\lambda$ is the solution to the quadratic equation

$$
\theta^2(-n\lambda') + \theta[(n - m + s)\lambda' - m] + s = 0.
$$

The roots x can be found with the quadratic formula, and $\tilde{\theta}_\lambda$ is the root satisfying the constraint $0 < x < 1$. $\tilde{\theta}_\lambda = s/m$ for $\lambda = 1.0$ and $\tilde{\theta}_\lambda = s/n$ for $\lambda = \infty$.

References

1. Vesikari T., Karvonen A., Korhonen T., Espo M., Lebacq E., Forster J., Zepp F., Delem A., De Vos B. Safety and immunogenicity of RIX4414 live attenuated human rotavirus vaccine in adults, toddlers and previously uninfected infants. *Vaccine*, 22, 2836–2842, 2004
2. Harro C.D., Pang Y.Y.S., Roden R.B.S., Hildesheim A., Wang Z., Reynolds M.J., Mast T.C., Robinson R., Murphy B.R., Karron R.A., Dillner J., Schiller J.T., Lowy D.R. Safety and immunogenicity trial in adult volunteers of a human papillomavirus 16 L1 virus-like particle vaccine. *Journal of the National Cancer Institute*, 93, 284–292, 2001
3. Feiring B., Fuglesang J., Oster P., Naess L.M., Helland O.S., Tilman S., Rosenqvist E., Bergsaker M.A.R., Nøkleby H., Aaberge I.S. Persisting immune responses indicating long-term protection after booster dose with meningococcal group B outer membrane vesicle vaccine. *Clinical and Vaccine Immunology*, 13, 790–796, 2006
4. Newcombe R.G. Two-sided confidence intervals for the single proportion: Comparison of seven methods. *Statistics in Medicine*, 17, 857–872, 1998
5. Borkowf C.B. Constructing binomial confidence intervals with near nominal coverage by adding a single imaginary failure or success. *Statistics in Medicine*, 25, 3679–3695, 2006
6. Lancaster H.O. Statistical control of counting experiments. *Biometrika*, 39, 419–422, 1952
7. Lancaster H.O. Significance tests in discrete distributions. *Journal of the American Statistical Association*, 56, 223–234, 1961
8. Miettinen O., Nurminen M. Comparative analysis of two rates. *Statistics in Medicine*, 4, 213–226, 1985
9. Chick S.E., Barth-Jones D.C., Koopman J.S. Bias reduction for risk ratio and vaccine effect estimators. *Statistics in Medicine*, 20, 1609–1624, 2001
10. Suissa S., Shuster J.J. Exact unconditional sample sizes for the 2×2 binomial trial. *Journal of the Royal Statistical Society*, Series A-Statistics in Society, 148, 317–327, 1985
11. Lydersen S., Fagerland M.W., Laake P. Tutorial in biostatistics: Recommended tests for association in 2×2 tables. *Statistics in Medicine*, 28, 1159–1175, 2009
12. Chan I.S.F. Exact tests of equivalence and efficacy with a non-zero lower bound for comparative studies. *Statistics in Medicine*, 17, 1403–1413, 1998
13. Berger R.L. Multiparameter hypothesis testing and acceptance sampling. *Technometrics*, 24, 295–300, 1982
14. Laska N.S., Tang D., Meisner M.J. Testing hypothesis about an identified treatment when there are multiple endpoints. *Journal of the American Statistical Association*, 87, 825–831, 1992
15. Chuang-Stein C., Stryszak P., Dmitrienko A., Offen W. Challenge of multiple co-primary endpoints: A new approach. *Statistics in Medicine*, 26, 1181–1192, 2007
16. Senn S., Bretz F. Power and sample size when multiple endpoints are considered. *Pharmaceutical Statistics*, 6, 161–170, 2007
17. Reed G.F., Meade B.D., Steinhoff M.C. The reverse cumulative distribution plot: A graphic method for exploratory analysis of antibody data. *Pediatrics*, 96, 600–603, 1995
18. Nauta J.J.P., Beyer W.E.P., Osterhaus A.D.M.E. On the relationship between mean antibody level, seroprotection and clinical protection from influenza. *Biologicals*, 37, 216–221, 2009

19. Lachenbruch P.A., Rida W., Kou J. Lot consistency as an equivalence problem. *Journal of Biopharmaceutical Statistics*, 14, 275–290, 2004
20. Julious S.A. *Sample Sizes for Clinical Trials*. Chapman and Hall/CRC, 2009
21. Fleiss J.L., Levin B., Paik M.C. *Statistical Methods for Rates and Proportions, Third Edition*. Wiley-InterScience, Hoboken, New Jersey, 2003
22. Hirji K.F., Tang M.L., Vollset S.E., Elashoff R.M. Efficient power computation for exact and mid-P tests for the common odds ratio in several 2x2 tables. *Statistics in Medicine*, 13, 1539–1549, 1994
23. Ting Lee M.L., Whitmore G.A. Statistical inference for serial dilution assay data. *Biometrics*, 55, 1215–1220, 1999
24. Nauta J.J.P., de Bruijn I.A. On the bias in HI titers and how to reduce it. *Vaccine*, 24, 6645–6646, 2006
25. Lyng J., Weis Bentzon M. International standards for anti-poliovirus sera types 1, 2 and 3. *Bulletin of the World Health Organization*, 29, 711–720, 1963
26. Nauta J.J.P. Eliminating bias in the estimation of the geometric mean of HI titers. *Biologicals*, 34, 183–186, 2006
27. Cox D.R., Oakes D. *Analysis of Survival data*. Chapman and Hall, London, 1984
28. Wang W.W.B., Mehrotra D.V., Chan I.S.F., Heyse J.F. Statistical considerations for nonInferiority/equivalence trials in vaccine development. *Journal of Biopharmaceutical Statistics*, 16, 429–441, 2006
29. Schuirmann D.J. A comparison of the two one-sided tests procedure and the power approach for assessing equivalence of average bioavailability. *Journal of Pharmacokinetics and Biopharmaceutics*, 15, 657–680, 1987
30. United States Food and Drug Administration. Guidance for Industry: Clinical Data Needed to Support the Licensure of Seasonal Inactivated Influenza Vaccines, May 2007
31. United States Food and Drug Administration. Guidance for industry: Clinical Data Needed to Support the Licensure of Pandemic Influenza Vaccines., May 2007
32. Joines R.W., Blatter M., Abraham B., Xie F., De Clercq N., Baine Y., Reisinger K.S., Kuhnen A., Parenti D.L. Prospective, randomized, comparative US trial of a combination hepatitis A and B vaccine, Twinrix. with corresponding monovalent vaccines, Havrix and Engerix-B. in adults. *Vaccine*, 19, 4710–4719, 2001
33. Nauta J. Statistical analysis of influenza vaccine lot consistency studies. *Journal of Biopharmaceutical Statistics*, 16, 443–452, 2006
34. ICH E9 Expert Working Group. Statistical Principles for Clinical Trials: ICH Harmonised Tripartite Guideline., September 1998
35. Wiens B.L., Iglewicz B. On testing equivalence of three populations. *Journal of Biopharmaceutical Statistics*, 9, 465–483, 1999
36. Kong L., Kohberger R.C., Koch G.G. Type I error and power in noninferiority/equivalence trials with correlated multiple endpoints: An example from vaccine development trials. *Journal of Biopharmaceutical Statistics*, 14, 893–907, 2004
37. Kong L., Kohberger R.C., Koch G.G. Design of vaccine equivalence/non-inferiority trials with correlated multiple binomial endpoints. *Journal of Biopharmaceutical Statistics*, 16, 555–572, 2006
38. European Agency for the Evaluation of Medicinal Products, Committee for Proprietary Medicinal Products, CPMP. Points to consider on switching between superiority and noninferiority. London, 27 July 2000
39. Dunnett C. W., Gent M. An alternative to the use of two-sided tests in clinical trials. *Statistics in Medicine*, 15, 1729–1738, 1976
40. Morikawa T., Yoshida M. A useful testing strategy in phase III trials: Combined test of superiority and test of equivalence. *Journal of Biopharmaceutical Statistics*, 5, 297–306, 1995
41. Ng T.-H. Simultaneous testing of noninferiority and superiority increases the false discovery rate. *Journal of Biopharmaceutical Statistics*, 17, 259–264, 2007
42. Leroux-Roels I., Vets E., Freese R., Seiberling M., Weber F., Salamand C., Leroux-Roels G. Seasonal influenza vaccine delivered by intradermal microinjection: A randomised controlled safety and immunogenicity trial in adults. *Vaccine*, 26, 6614–6619, 2008

43. Ganju J., Izu A., Anemona A. Sample size for equivalence trials: A case study from a vaccine lot consistency study. *Statistics in Medicine*, 27, 3743–3754, 2008
44. Kohberger R.C. Comments on Sample size for equivalence trials: A case study from a vaccine lot consistency trial by J. Ganju, A. Izu and A. Anemona. *Statistics in Medicine*, 28, 177–178, 2009
45. Ganju J., Izu A., Anemona A. Authors' Reply. *Statistics in Medicine*, 28, 178–179, 2009
46. Halloran M.E., Longini I.M., Struchiner C.J. Design and interpretation of vaccine field studies. *Epidemiological Reviews*, 21, 73–88, 1999
47. Halloran M.E. Overview of vaccine field studies: Types of effects and designs. *Journal of Biopharmaceutical Statistics*, 16, 415–427, 2006
48. Halloran M.E., Longini, Jr. I.M., Struchiner C.J. *Design and Analysis of Vaccine Studies*. Springer, New York, 2009
49. Orenstein W.A., Bernier R.H., Hinman A.R. Assessing vaccine efficacy in the field - Further observations. *Epidemiological Review*, 10, 212–241, 1988
50. Blennow M., Olin P., Granström M., Bernier R.H. Protective efficacy of a whole cell pertussis vaccine. *British Medical Journal*, 296, 1570–1572, 1988
51. Urdaneta M., Prata A., Struchiner C.J., Tosta C.E, Tauil P., Boulos M. Evaluation of SPf66 malaria vaccine efficacy in Brazil. *American Journal of Tropical Medicine and Hygiene*, 58, 378–385, 1988
52. Moorthy V., Reed Z., Smith P.G. Measurement of malaria vaccine efficacy in phase III trials: Report of a WHO consultation. *Vaccine*, 25, 5115–5123, 2007
53. Jahn-Eimermacher A., Du Prel J.B., Schmitt H.J. Assessing vaccine efficacy for the prevention of acute otitis media by pneumococcal vaccination in children: A methodological overview of statistical practice in randomized controlled clinical trials. *Vaccine*, 25, 6237–6244, 2007
54. McCullagh P., Nelder J.A. *Generalized Linear Models*, Second Edition. Chapman and Hall, London, 1989
55. Andersen P.K., Gill R.D. Cox regression model for counting processes: A large sample study. *Annals of Statistics*, 10, 1100–1120, 1982
56. Forrest B.D., Pride M.W., Dunning A.J., Capeding M.R., Chotpitayasunondh T., Tam J.S., Rappaport R., Eldridge J.H., Gruber W.C. Correlation of cellular immune responses with protection against culture-confirmed influenza virus in young children. *Clinical and Vaccine Immunology*, 15, 1042–1053, 2008
57. Siber G.R. Methods for estimating serological correlates of protection. *Developments in Biological Standardization*, 89, 283–296, 1997
58. Hobson B., Curry R.L., Beare A.S., Ward-Gardner A. he role of serum haemagglutination-inhibiting antibody in protection against challenge infection with influenza A2 and B viruses. *Journal of Hygiene*, 70, 767–777, 1972
59. Dunning A.J. A model for immunological correlates of protection. *Statistics in Medicine*, 25, 1485–1497, 2006
60. White C.J., Kuter B.J., Ngai A., Hildebrand C.S., Isganitis K.L., Patterson C.M., Capra A., Miller W.J., Krah D.L., Provost P.J., Ellis R.W., Calandra G.B. Modified cases of chickenpox after varicella vaccination: correlation of protection with antibody response. *Pediatric Infectious Disease Journal*, 11, 19–23, 1992
61. Jdar L., Butler J., Carlone G., Dagan R., Goldblatt D., Kyhty H., Klugman K., Plikaytis B., Siber G., Kohberger R., Chang I., Cherian T. Serological criteria for evaluation and licensure of new pneumococcal conjugate vaccine formulations for use in infants. *Vaccine*, 21, 3265–3272, 2003
62. Ellenberg S.S. Safety considerations for new vaccine development. *Pharmacoepidemiology and Drug Safety*, 10, 411–415, 2001
63. Mutsch M., Zhou W., Rhodes P., Bopp M., Chen R.T., Linder T., Spyr C., Steffen R. Use of the inactivated intranasal influenza vaccine and the risk of Bell's palsy in Switzerland. *The New England Journal of Medicine*, 350, 896–903, 2004
64. Whitaker H.J., Farrington C.P., Spiessen B., Musonda P. Tutorial in biostatistics: The self-controlled case series method. *Statistics in Medicine*, 25, 1768–1798, 2006

65. Dmitrienko A., Tamhane A.C., Bretz F. (editors) *Multiple Testing Problems in Pharmaceutical Statistics*. Chapman and Hall/CRC, New York, 2010

66. Benjamini Y., Hochberg Y. Controlling the false discovery rate: A practical and powerful approach to multiple testing. *Journal of the Royal Statistical Society*, Series B 1995, 57, 289–300, 1995

67. Mehrotra D.V., Heyse J.F. Use of the false discovery rate for evaluating clinical safety data. *Statistical Methods in Medical Research*, 13, 227–238, 2004

68. Berry S.M., Berry D.A. Accounting for multiplicities in assessing drug safety: A three-level hierarchical mixture model. *Biometrics*, 60, 418–426, 2004

69. Singh M., ed. *Vaccine Adjuvants and Delivery Systems*. Wiley, Hobokon, New Jersey, 2007

70. United States Food and Drug Administration. Guidance for Industry: Toxicity Grading Scale for Healthy Adults and Adolescent Volunteers enrolled in Preventive Vaccine Clinical Trials., September 2007

71. De Bruijn I.A., Nauta J., Gerez L., Palache A.M. The virosomal influenza vaccine Invivac: Immunogenicity and tolerability compared to an adjuvanted influenza vaccine, Fluad. in elderly subjects. *Vaccine*, 24, 6629–6631, 2006

72. Lehmann E.H.J.M. *Nonparametrics: Methods Based on Ranks*. Springer Science+Bussiness Media, LLC, New York, USA, 2006

73. Diggle L., Deeks J. Effect of needle length on incidence of local reactions to routine immunisation in infants aged 4 months: randomised controlled trial. *British Medical Journal*, 321, 931–933, 2000

Index